Karl-Heinz Dieser

Optimizing the sustainability of European freight transport
- Green Energy in the mobility competition -

Karl-Heinz Dieser

Optimizing the sustainability of European freight transport

- Green Energy in the mobility competition -

Impressum

Bibliografische Information der Deutschen Nationalbibliothek: Die Deutsche Nationalbibliothek verzeichnet diese Publikation in der Deutschen Nationalbibliografie; detaillierte bibliografische Daten sind im Internet über http://dnb.dnb.de abrufbar.

Lektorat: Vorname Nachname oder Institution
Korrektorat: Vorname Nachname oder Institution
Weitere Mitwirkende: Vorname Nachname oder Institution

Verlag: BoD · Books on Demand GmbH, Überseering 33, 22297 Hamburg, bod@bod.de

Druck: Libri Plureos GmbH, Friedensallee 273, 22763 Hamburg

ISBN: 978-3-6963-9174-4

MIX
Papier aus verantwortungsvollen Quellen
Paper from responsible sources
FSC® C105338

Content

LIST OF ABBREVIATIONS VIII

LIST OF FIGURES X

LIST OF TABLES XI

ABSTRACT XII

1 INTRODUCTION 13

1.1 Relevance and topicality of the topic 14

1.2 Current state of research 17

1.3 Objectives and research questions 18

1.4 Structure of the work 20

2 CONCEPTUAL FOUNDATIONS 22

2.1 Green Energy 22

2.1.1 Definition of Green Energy 22

2.1.2 European Green Deal 23

2.1.3 European climate targets 24

2.2 Sustainability 25

2.2.1 Concept of sustainability 25

2.2.2 Forms of sustainability 25

2.2.3 Ecological view of sustainability 27

2.2.4 Economic view of sustainability 28

2.2.5 Social view of sustainability 29

2.2.6 Technology as the fourth dimension of sustainability 30

2.3 Mobility 31

2.3.1 Concept of mobility 31

2.3.2 Forms of mobility 32

2.3.3 Definition of the concept of mobility 32

2.3.4 Mobility in freight transport 33

2.3.5 Sustainable mobility 34

2.3.6 Mobility competition in Europe 34

2.4 Freight transport 35

2.4.1 Definition of freight transport 35

2.4.2 Forms of freight transport 36

2.4.3 Differentiation between freight transport and logistics 37

2.4.4 The state of green energy in freight transport 37

2.5 Interim conclusion 39

3 FREIGHT TRANSPORT IN EUROPE 41

3.1 Mobility in the transportation sector 43
3.1.1 Energy consumption and emissions 45
3.1.2 Infrastructure expansion 46
3.1.3 Cargo tubes 47
3.2 Freight transport in Germany 48
3.3 Freight transport in Europe 51
3.4 Differentiation from intermodal maritime transport 53
3.5 Interim conclusion 54
4 RESEARCH DESIGN AND EMPIRICAL VALIDATION 57
4.1 Literature analysis as the basis for the design 58
4.2 Presentation of the SWOT analysis 59
4.3 Derivation of the causal model 60
4.4 Empirical Investigation 62
4.4.1 Derivation of the research questions 63
4.4.2 Quality criteria for scientific research 64
4.4.3 Selection of the methodology 65
4.5 Presentation of Arena analysis 66
4.6 Presentation of expert interviews 69
4.6.1 Interview planning 69
4.6.2 Creation of the interview guide 71
4.6.3 Implementation 75
4.7 Transcription and evaluation 76
4.7.1 Generalization, paraphrasing and condensation 77
4.7.2 Quality criteria according to Mayring 78
4.7.3 Potential for Bias 79
4.7.4 Optimization according to the 7S Model 80
5 IMPLEMENTATION AND EVALUATION 81
5.1 Results of the arena analysis 81
5.2 Results of the SWOT analysis 84
5.3 Results of the expert interviews 88
5.4 Review of the quality criteria 96
5.5 Derivation of optimization recommendations 98
5.6 Interim conclusion 99
6 CONCLUSION 101
6.1 Answering the research questions 101
6.2 Limitation of work 107

6.3 Implication for science 108
6.4 Optimization recommendation for freight transport 109
6.5 Outlook 111

LITERATURE 114

APPENDIX 136

Appendix 1: AI tools directory 136
Appendix 2: Word Cloud 137
Appendix 3: Interview guide 138
Appendix 4: Structure of the work 140
Appendix 5: List of the expert interviews conducted 141
Appendix 6: Deductive category system 142
Appendix 7: Coding guide and coding rules 143
Appendix 8: Paraphrasing and reduction I of transcripts 144
Appendix 9: Generalization and reduction II of transcripts 172
Appendix 10: 4 x 3 matrix 195
Appendix 11: Literature search table 196
Appendix 12: PRISMA Flow Diagram 198
Appendix 13: Model research table 199
Appendix 14: Model synthesis table 226
Appendix 15: Research poster (Draft) 239
Appendix 16: Research poster (English version) 240
Appendix 17: Research poster (German version) 241
Appendix 18: Research poster (Bulgarian version) 242
Appendix 19: Summarizing the SWOT analysis 243
Appendix 20: Time schedule 244
Appendix 21: Written validation of the transcripts 245
Appendix 22: Articles by Author 246

KEYWORD INDEX 247

List of abbreviations

°C	Degree Celsius
AI	Artificial intelligence
BMWK	Bundesministerium für Wirtschaft und Klima
Cat.	Category
CEO	Chief Executive Officer
CEP	Courier, Express and Parcel services
CFO	Chief Financial Officer
CGI	Consultants to Government and Industry
CIO	Chief Information Officer, IT Manager
CM	Cargo Management
CO_2	Carbon dioxide
CSM	Center for Sustainability Management
CSR	Corporate Social Responsibility
DAC	Digital Automatic Coupling
DB	German Railways (Deutsche Bahn)
DHL	Dalsey, Hillbrom and Lynn
DVZ	German Transport Newspaper (Deutsche Verkehrs Zeitung)
e.g.	For example
EEA	European Environmental Agency
ESG	European Social Governance
ETCS	European Train Control System
EU	European Union
GHG	Greenhouse gas
HGV	Heavy Goods Vehicle
kg	kilogram
KOFA	Competence center for securing skilled workers
KSG	Climate Protection Act
kWh	Kilowatt hour
LoE	Level of Evidence
MD	Managing Director
ÖBB	Austrian Federal Railways (Österreichische Bundes Bahn)
OEM	Original Equipment Manufacturer
PJ	Peta Joule
ppm	Parts per million stand for a factor of one millionth
RoI	Return on investment
SBB	Swiss Federal Railways

t	Ton
tkm	Metric ton kilometer
UBA	Environment Federal Office
URL	Uniform Resource Locator
VCD	Verkehrsclub Deutschland
VDV	Association of German Transport Companies

LIST OF FIGURES

Figure 1: Connection of the research areas 14
Figure 2: Greenhouse gas trend until 2050 16
Figure 3: Four pillars of sustainability 26
Figure 4: Forms of freight transport 36
Figure 5: EU Single Market 42
Figure 6: Freight transport performance by mode of transport in Germany 49
Figure 7: Ecological consequences Maritime freight 54
Figure 8: Qualitative evaluation method 57
Figure 9: Template SWOT analysis 59
Figure 10: Causal model 61
Figure 11: Principles of research 65
Figure 12: Template of the extended environment analysis 68
Figure 13: Planning the interviews 69
Figure 14: Quality criteria according to Mayring 79
Figure 15: Application of the extended environment analysis 83
Figure 16: SWOT analysis - strengths and weaknesses 84
Figure 17: SWOT analysis - opportunities and risks 86
Figure 18: Word Cloud 137
Figure 19: Strukture of the work 140
Figure 20: 4 x 3 matrix 195
Figure 21: PRISMA Flow diagram 198
Figure 22: Research Poster 239
Figure 23: Research poster in English language 240
Figure 24: Research poster in Garman language 241
Figure 25: Research poster in Bulgarian language 242
Figure 26: Summarization of the SWOT analysis 243
Figure 27: Time schedule 244

LIST OF TABLES

Table 1: Transport performance within Germany 44
Table 2: Energy consumption by mode of transport 46
Table 3: Traffic volume and transport performance in freight transport 50
Table 4: Stakeholders and issues in European freight transport 67
Table 5: Categorization of the identified stakeholders 67
Table 6: Questions for the introductory phase 72
Table 7: Questions on potentials and risks 73
Table 8: Questions on environmental sustainability 73
Table 9: Questions on social sustainability 73
Table 10: Questions on economic sustainability 74
Table 11: Questions on technical sustainability 74
Table 12: Green Energy 75
Table 13: Derivation to questions of the final question + other topics 75
Table 14: List of experts for interviews 76
Table 15: Derivation of the deductive categories 87
Table 16: Recommendation for action in the 7S model according to McKinsey 98
Table 17: Proof of the expert interviews conducted 141
Table 18: Category system 142
Table 19: Coding guide and coding rules 143
Table 20: Paraphrasing and reduction I 171
Table 21: Generalization and reduction II 194
Table 22: Literature search table 197
Table 23: Model research table 225
Table 24: Model Synthesis Table 238

ABSTRACT

In 2023, the EU states' climate and energy targets for 2040 were adjusted as part of the Fit for 55 packages to meet the EU's ambitions. The climate targets for 2030 require a significant reduction in CO_2 emissions in freight transport in order not to exceed 85 million tons of CO_2. Although some progress has been made, no clear trend reversal is observed. Private transport poses a particular challenge, which is why intermodal transport solutions and intelligent transport systems are seen as key approaches to reducing environmental pollution.

Stricter emission limits have applied from 2023 and EU member states had to adapt their energy and climate plans. Forecasts predict a 45% increase in freight transport by 2040, which could lead to higher emissions if no countermeasures are taken. Technological innovations, such as the Internet of Things and big data, as well as renewable energy are crucial to achieving the climate targets.

Freight transport continues to be heavily dominated by road traffic, which requires additional measures to reduce emissions. A shift to more environmentally friendly modes of transport such as rail and the expansion of infrastructure, including alternative drive systems such as hydrogen and electrification, are necessary. International cooperation and standardized approaches are also important to promote the sustainability of European freight transport.

Climate change requires an integrated approach that combines technological innovations and political measures to promote sustainable transport systems and reduce CO_2 emissions in freight transport.

1 Introduction

Alongside industry and the energy sector, freight transport is the third largest producer of greenhouse gases (cf. European Parliament 2024). Electricity generation, heat generation and transport activities have a major impact on the environment due to greenhouse gas emissions (cf. Umweltbundesamt2024). Greenhouse gases are mainly released through the use of fossil fuels such as coal or oil. Finally energy consumption in freight transport is at a high level. It will therefore be difficult to achieve the environmental targets that have been set (cf. Umweltbundesamt 2024).

The European Environmental Agency (EEA) illustrated the development of freight transport as early as 2023 (cf. Environmental Agency 2023). Freight transport is characterized by the fact that it is the only economic sector in which CO_2 emissions continue to rise. This has already been established by Schwedes (2019: 114). Freight transport is therefore still not contributing to climate protection. Greenhouse gas emissions have remained almost the same in recent years. Freight transport has become significantly more efficient since the early 1990s, but the increase in transport performance has led to an increase in final energy consumption (cf. Umweltbundesamt 2024). According to the EEA (2024), estimates indicate that the share of renewable energy sources has increased in many member states of the European Union (EU) and that energy consumption has been reduced.

The use of innovative transportation solutions can reduce environmental pollution and the associated traffic congestion (cf. Ibasha & Abdellatif 2025). A reduction in emissions can be achieved with innovative transportation solutions (cf. Fischer 2024). Sustainable transport solutions and innovative drives are key elements of climate-friendly freight transport (cf. Schmitz 2023). According to Schneider (2024: 41), an increase in efficiency contributes to a reduction in environmental pollution. Decarbonization of freight transport is the key to achieving global climate targets (cf. Browne 2023a). Increasing efficiency and reducing emissions are therefore key objectives of sustainable mobility in European freight transport (cf. Grant 2024). One trend is the development of sustainable logistics concepts through innovative products such as artificial intelligence (AI) (cf. Weber 2023). AI is developing at a rapid pace and is helping to increase sustainability (cf. Wensauer 2023). AI thus ensures the development of logistics concepts.

These considerations result in a research area that is examined in more detail in this dissertation. Figure 1 shows the links between the different research areas of this dissertation. Essentially, the following three sub-areas are considered: *freight transport in Europe*, the *sustainability megatrend* and *green energy*. These three research areas give rise to further topics that will be considered over time.

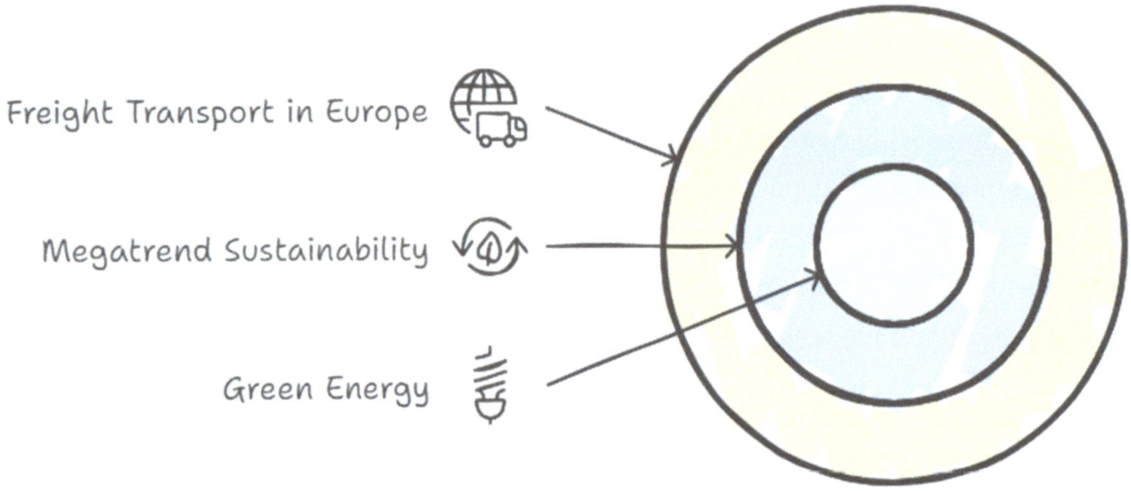

Figure 1: Connection of the research areas
Source: Author's own representation.

In this context, the coverage of the digital infrastructure in Europe is considered with the megatrend of connectivity. Without the advancement of digitalization, connectivity and mobility cannot be further expanded. Innovative technologies such as autonomous driving, AI and telematics systems play a key role in sustainable mobility (cf. Cullinane 2024). Technological change is generally one of the relevant factors without which implementation is not possible. Innovative technologies such as electric vehicles or hydrogen drives are key aspects in the promotion of sustainable freight transport (cf. Browne 2023b). Sustainability is playing an increasingly important role in freight transport. The trend is moving away from the combustion of fossil fuels and towards renewable energy sources and drives (cf. Ember 2025). Germany has set a goal to achieve zero emissions by 2045. Some cities aim for climate neutrality five years ahead of schedule. A decisive step on the way to achieving this is the expansion of renewable energy (cf. Tagesschau 2024b). Current areas of research are focusing on the production of green energy and the corresponding drives. "It is fundamentally positive that renewable energy is being expanded more quickly than in the past," says Claudia Kemfert, energy expert at the German Institute for Economic Research (2024). However, this development is no reason for optimism, as expansion remains too slow to meet the 2045 climate goals.

1.1 Relevance and topicality of the topic

In 2023, the climate and energy targets of the 27 EU member states for 2040 were revised as part of the Fit for 55 package. The climate targets were adjusted to align with the updated European targets (cf. European Environmental Agency 2023). In order to achieve the climate targets set, a significant reduction in emissions is necessary. According to the Federal Climate Protection Act (2022), the target for freight transport is 85 million tons of CO_2 in 2030. The Federal Government wants to reduce final energy consumption in freight and passenger transport by 15 to 20 % by 2030 compared to 2005 (cf. Umweltbundesamt2024).

Meeting climate targets is further complicated by the emphasis on private transport (cf. Saake & Sommer 2023). The key aspect of sustainability can be achieved by reducing CO_2 emissions (cf. Jackson 2023a: 92). The European targets have been adjusted, with both higher targets for 2040 and stricter emission limits from 2023. Member states were also required to submit a draft update of their climate plans reflecting the revised EU climate and energy targets for 2040 (cf. European Environmental Agency 2023). Transport systems are already regularly reaching the limits of their capacity. The consequences are traffic noise and air pollution. In order to counteract the overloading of transport capacities, data analyses are used to optimize transport routes (cf. Internationale Wochenzeitung 2024). According to Chiara (2023), this results in intelligent transportation systems that can increase efficiency and reduce environmental pollution. Intelligent transport systems and data analysis are crucial for optimizing sustainable mobility solutions. It is predicted that freight traffic will continue to increase in the coming years. Total freight traffic in Europe is expected to increase by around 45%.

According to Wagner (2023: 98), transport volumes are expected to continue to rise in the coming years. This means that CO_2 emissions are also expected to rise. The use of renewable energy is crucial for climate-friendly freight transport. Freight transport has a role to play in climate change that should not be neglected. The increasing growth in freight transport is countered by climate targets, as the goal of reducing CO_2 emissions by up to 55% is to be achieved in the same period (cf. Umweltbundesamt2024). The integration of sustainability criteria into transport planning is increasing. These can be explained as follows:

- **Freight transport trends:** Projections for both road and rail freight transport suggest an increase in transport volumes through 2040.
- **Road transport is growing faster than rail freight,** showing increased reliance on roads for freight.
- **Impact on the environment**: An increase in total freight transport, particularly road transport, may result in higher CO2 emissions, which does not align with established climate targets aimed at reducing emissions (cf. Jackson 2023a: 103).

The trend towards decarbonization requires the use of innovative technologies and new business models (cf. Schmitz 2023). These technologies are driving the efficiency of freight transport (cf. Cole 2023). As a result, the climate targets below can be used as an opportunity to analyze the European transport market. After examining competition and mobility, it is possible to identify optimization recommendations. The climate targets of the European Commission are (cf. Federal Statistical Office 2023):

- Increase in traffic in Europe by 45% by 2050
- Reduction of CO2 emissions by 55% by 2050

These two climate targets are illustrated in Figure 2, which shows the transport sector in general. It has a non-negligible role to play in climate change (cf. European Commission 2024a). This scientific paper therefore examines the topic of sustainability with a focus on European freight transport. The figures show that the issue of sustainable mobility is of great concern to the population and that these concerns will continue to grow.

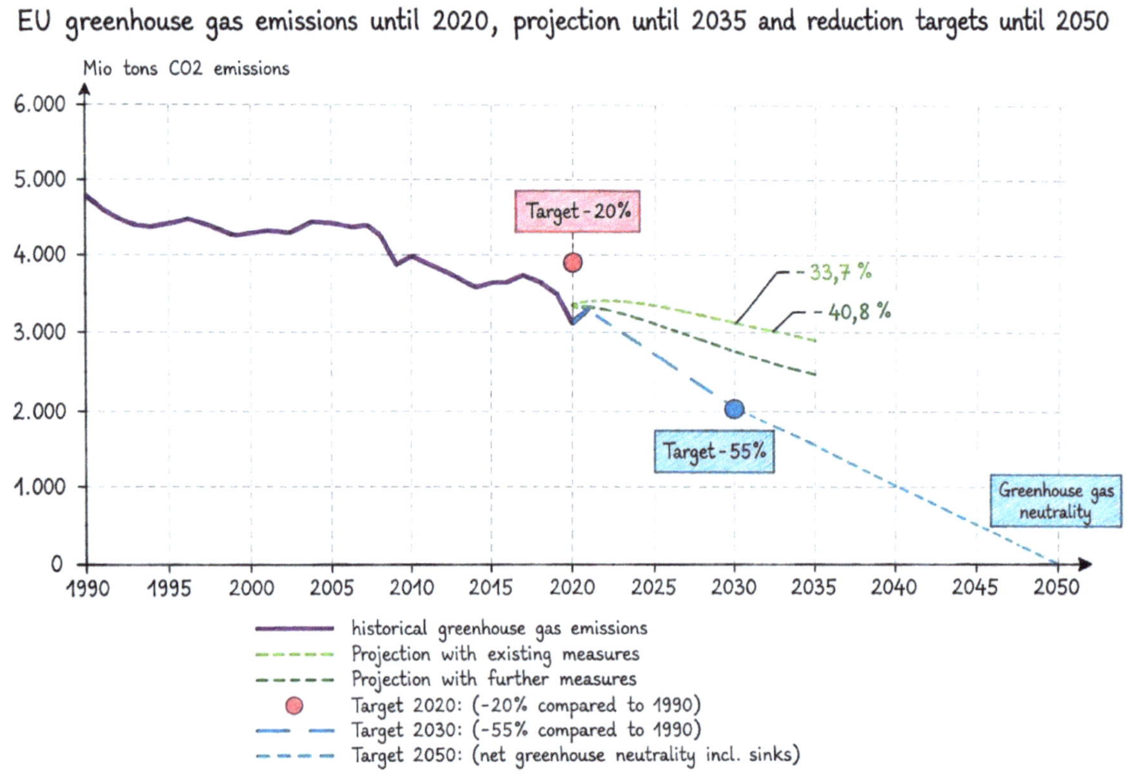

Figure 2: Greenhouse gas trend until 2050
Source: Umweltbundesamt (2024).

In 2023, new laws were introduced that will continue to impact society. European policy has significantly increased the complexity of the framework conditions for freight transport. The decarbonization of energy has gained considerable momentum in recent years. In 2023, the share of renewable energy rose to over 50 percent for the first time (cf. Ernest & Young 2024). In addition to financing and GHG quotas, demands for energy savings and management systems are other topics that were discussed in 2023 (cf. VDV 2024). According to the Federal Statistical Office (2021), these trends and the associated political decisions will contribute significantly to the fact that road transport will continue to grow, while the shift to more environmentally friendly modes of transport such as rail is progressing more slowly.

Over 1.3 million trucks drive on German highways every day. Around a third come from European countries (cf. Verkehrsclub 2023). Rail companies have to pay for the use of the route they use. However, there are strict restrictions on the charging of tolls on highways. However, this is not rail companies' only disadvantage compared to road transport. Furthermore, European railroads continue to operate with varying signalling systems and track gauges, as standardisation has not yet been achieved. Another aspect is that the UN projects that the population will grow to around 9.7 billion people by 2050 at the current rate of growth. This corresponds to an increase of 25% over the next 25 years. It can be assumed that this rapid population growth will be subject to strong regional fluctuations. This may result in a sharp increase in urbanization. According to Siebenpfeiffer (2021), it will increase from 55% to 68% by 2050. These advancements create significant requirements for the freight transport system. Congestion on the highways is already a daily occurrence. According to the Association of German Engineers (2024), rail and inland waterway vessels are the most efficient and environmentally friendly means of transport They produce significantly less CO_2 per metric ton kilometer than road and air transport (cf. Carter 2023).

1.2 Current state of research

The current state of scientific knowledge clearly shows that the transition to sustainable energy and transport systems is a key requirement for achieving international climate goals. Central to this are green energy, sustainability in its various dimensions, and sustainable mobility and freight transport concepts. Green energy comes from natural, renewable sources such as solar, wind, or hydro power and causes little or no CO_2 emissions. It must be distinguished from the terms renewable and clean energy. While renewable energy refers to the source, clean energy is low emission but not necessarily nature based.

Despite increasing political efforts, such as those set out in the European Green Deal, several challenges remain, including corporate greenwashing, technological barriers, high investment costs, and ecological concerns related to energy production. The Green Deal aims to reduce emissions by 55% by 2030 and achieve climate neutrality by 2050, in line with the Paris Climate Agreement.

The concept of sustainability has expanded to include a technological dimension. In addition to the traditional ecological, economic, and social pillars, technological sustainability is becoming increasingly important. Modern technologies such as artificial intelligence, digitalization, and blockchain are seen as key enablers of sustainable development, especially in logistics and freight transport.

Freight transport contributes significantly to global emissions, around 70% of freight in Europe is still transported by road. The shift toward more sustainable modes such as rail, inland waterways, or low-emission vehicles is progressing slowly. Nevertheless, initiatives involving cargo tube systems, battery-electric and hydrogen-powered vehicles, alongside multimodal transport concepts, demonstrate significant potential to lower emissions and enhance operational efficiency.

Green logistics is gaining importance. The use of emission-free technologies and digital tools supports resource conservation and cost reduction. Still transforming the transport sector is a multifaceted process that involves coordinated actions across politics, industry, technology, and society. Sustainable energy and mobility are not just visions for the future but a necessity. Their implementation is not only ecologically essential but also economically sensible and crucial for Europe's competitiveness in the 21st century.

1.3 Objectives and research questions

One of the aims of this dissertation is to detail the current situation in the freight transport industry in Europe and how mobility is changing. It also aims to clarify where transport companies and politicians in Europe need to take action in order to further promote sustainability and achieve climate targets.

Another goal is to use literature and empirical research in an arena analysis to select suitable experts for interviews. The possible strengths and weaknesses resulting from the analysis are addressed and analyzed in the SWOT analysis and discussed in a separate chapter. The future-oriented opportunities and risks are based on the expert interviews. The results of the dissertation should serve as a basis for further investigations and for deriving further measures.

The overarching goal is to identify and derive optimization measures, which are presented in the form of recommendations for action. These actions are recommended following an evaluation of the current situation and future considerations, which are informed by insights from expert interviews that provide a forward-looking perspective. The McKinsey analysis is used to identify the most important measures.

The topic of sustainability is becoming increasingly relevant, as the megatrend of climate protection is being addressed alongside the megatrend of mobility. The volume of traffic will continue to increase in the coming years, while at the same time energy targets must be met with green energy. Green energy is increasingly coming to the fore. In order to bring all aspects to a positive conclusion, the market and its requirements must be analyzed in order to formulate valid optimization recommendations. Most research topics relate to electric mobility in private transportation by car. There is a research gap here that combines mobility, environmental aspects, and green energy. Another research gap is the technological side, which has not been sufficiently considered in recent years. A further argument is that these findings should be analyzed for European freight transport on the basis of current CO_2 emissions and developments, and green energy strategies should be derived from the results. The following central research question arises from these considerations. In addition to the core issue of mobility and its competition in Europe, it also covers sustainability issues and the associated environmental aspects.

Key question

What are the potentials and risks of mobility competition?

To address the main research question and develop recommendations for improving mobility competition, three additional sub-questions are formulated.

Sub-questions

(1) To what extent will European freight transport have to adapt in order to achieve the sustainability goals by 2040?

(2) What actions can be implemented to make European freight transport more sustainable in accordance with established climate targets?

(3) How can green energy be used sensibly in European freight transport to achieve sustainability goals?

To address these research questions, the author must take several basic steps. Efficient research is a prerequisite. Modern and efficient technology must be used for research projects. Scientific research involves a critical evaluation of the credibility and relevance of sources to ensure that only trustworthy information is used in the research. Accurate documentation and follow-up of results, along with formulating appropriate recommendations are crucial. These steps are ensured by the author of the dissertation.

1.4 Structure of the work

The structure of the thesis is visualized in the following figure. The appendices are not shown in the figure below.

Chapter 2 first explains the definition of green energy and how it differs from clean energy. This is followed by sustainability aspects such as social, economic dimensions and environmental dimensions. As the fourth and new aspect, technology is presented and the characteristics and special features of sustainability are pointed out. The following chapter discusses the concept of mobility and related topics. Sustainable mobility follows directly. An interim conclusion concludes this chapter.

This is followed by Chapter 3, which describes freight transport and energy consumption. European freight transport and current consumption and emissions during operation are presented. This is followed by a description of the current status of green energy in freight transport. Combined transport and cargo tubes are explained as key sub-items. The SWOT analysis, which presents the strengths and weaknesses of European freight transport, provides an indication of the current situation. This chapter ends with an interim conclusion.

The research design is explained in detail in chapter 4. The research design, including the quality criteria for scientific research and the choice of methodology, are explained in detail. The qualitative research process involves conducting an upstream arena analysis prior to expert interviews to identify relevant stakeholders. The planning of the expert interviews and the guidelines are explained, followed by a detailed discussion of the evaluation. In order to make the results more usable, the second part of the SWOT analysis is carried out to identify future opportunities and risks.

Chapter 5 deals with the implementation and evaluation of the analyses. The results of the arena analysis are followed by the results of the expert interviews with regard to the sustainability aspects of European freight transport. In the SWOT analysis, the potential and risks are analyzed. The chapter concludes with an interim conclusion.

Chapter 6 begins by answering the central research question and the three sub-questions, followed by a discussion of further findings. Chapter 6 concludes with final consideration. Next is a critical appraisal and its scientific implications. The subsequent limitation shows the restrictions. This is followed by possible solutions and optimization recommendations on how the climate targets in Europe can be achieved. Various approaches are explained.

The outlook for the future concludes Chapter 6 and thus also the thesis. Finally, there are various appendices such as the category system, the interview guide, various statistics, and templates that are relevant to this work. A summary of the expert interviews is attached.

The work comprises 31 annexes and begins with a presentation of transport volumes and performance in European freight transport. This is followed by a word cloud providing a keyword overview of the contents of the paper. Next are the interview guide and survey invitation. The following section provides a list of expert interviews and their corresponding transcripts. After the presentation of the deductive category system, the coding rules are described. The transcripts are then systematically condensed through paraphrasing (reduction 1) and generalization (reduction 2). A Gantt chart visualizes the time schedule for the years 2024 and 2025. A 4x3 matrix is then presented, followed by a table for the literature search, which is supplemented by a PRISMA flow chart. This diagram illustrates the literature search methodology. A research poster forms the transition to the condensation of the SWOT analysis. The selection of experts is also described in detail. Finally, the written validation of the transcripts by the participants in the expert interviews is documented.

2 Conceptual foundations

The following chapter provides a theoretical classification of the terms, contexts and models associated with the topic. The definition of green energy and sustainability valid for this work is outlined with all its tasks. Next is an overview of mobility and freight transport. An interim conclusion summarizes the environment.

2.1 Green Energy

Green energy is becoming increasingly common in times of climate change and the decarbonization of energy in Europe (cf. Crawford 2024). Green energy is first defined and dıfferentiated.

2.1.1 Definition of Green Energy

Green energy refers to energy produced without emitting CO_2 or other greenhouse gases (cf. EWE 2024.). Despite its CO_2 neutrality, nuclear energy is often viewed critically, as its construction, transportation and final storage of fuel rods have an impact on ecosystems (cf. Öko-Institut 2024).

In evaluating the environmental friendliness of green energy, it is important to differentiate between CO_2 emissions and other environmental impacts. Renewable energy is defined by the resources used, as these must be renewable in a natural way. Green energy is becoming increasingly attractive (cf. Euramco 2024). Many freight companies worry that they seem more eco-friendly to outsiders than they actually are (cf. The Sustainable Agency 2025.). Climate change is constantly driving forward technical issues in the field of clean energy. Efforts are underway to develop new technologies to meet United Nations climate targets outlined in the European Green Deal (cf. Financial Times 2024). These sustainability goals are set out in the Paris Climate Agreement. The participating countries have committed themselves to reducing global warming to below 2°C. (cf. Umweltbundesamt2024c). Differentiation between clean, green, and renewable energy

Green energy is renewable energy sourced from nature, such as water, wind, or sunlight. A distinction must be made between green energy and renewable energy (cf. Climate Portal 2024). Renewable energy is generated from recyclable sources, while green energy refers to energy from natural sources (cf. Euramco 2024). Green energy is distinct from clean and renewable energy (cf. Palmetto 2025.). Commonly used interchangeably in daily speech. Theoretically, renewable energy is inexhaustible because it is a continuous source of energy. However, renewable energy is not always clean, and this cleanliness depends on the EU countries' clean energy laws. Some renewable energy sources emit pollutants without exceeding the legal definition of *cleanliness.* The cleanest renewable energy sources are wind energy, geothermal energy, and solar energy. Although small hydropower plants are considered renewable and clean, large hydropower facilities may release certain pollutants and pose environmental risks during their construction. They therefore generate renewable but not clean energy (cf. Palmetto 2025).

Clean energy causes little or no emissions during its generation and use. Green energy sources release small amounts of pollutants, often during generation and use. Green energy production emits fewer greenhouse gases than fossil fuels, though it is not completely clean. Due to the small amounts of greenhouse gases produced during their use, renewable power does not threaten habitat loss (cf. Dalrada 2024). These energy sources do not cause any significant emissions during use and production and are renewable. Wind and sun are clean and renewable energy sources. Their supply cannot be depleted, as the sun always provides solar energy when it produces light and the wind always turns wind turbines when it blows to generate energy (cf. Climate+ Portal 2024).

Terms like green, clean, or renewable energy refer to sources that lessen reliance on fossil fuels (cf. BMWK 2024a). However, they do not all have the same meaning when it comes to how energy is produced, used and regenerated (cf. BMWK 2024). While these terms have minor distinctions, they indicate the direction the energy company is taking regarding climate change (cf. Palmetto 2025). Since no single sustainable energy source can meet all energy needs, a combination of different green, clean and renewable energy sources is necessary (cf. First climate 2024).

2.1.2 European Green Deal

In 2021, the European Parliament passed the EU Climate Law, requiring a 55% emissions reduction by 2030 and climate neutrality by 2050 (cf. European Council 2024). In February 2024, the European Commission formulated the goal of reducing greenhouse gas emissions by 90 % by 2040 to achieve climate neutrality by 2050 (cf. European Commission 2024). This moves the EU nearer to its 2050 climate neutrality targets. European Commission 2024c). The benefits are cleaner air and cleaner water. The economy will also benefit, as new standards will be developed. The creation of jobs and new professional fields for energy-efficient buildings and processes are further advantages (cf. European Parliament 2023). The European Green Deal outlines key climate policy goals focused on achieving climate neutrality in the future (cf. Umweltbundesamt 2023).

According to the European Commission, climate change is one of the greatest challenges and offers an opportunity to develop new economic models. The European Green Deal is the basis for the energy transition (cf. Bothe 2024). This energy transition offers many benefits, starting with new opportunities for innovation, investment and green jobs. All EU Member States have committed to making the EU a climate-neutral continent by 2050. As a step towards climate neutrality, the European Commission set a further climate target for 2040 in February 2024 (cf. Association of Local Authorities 2024). It recommended a 90% reduction in greenhouse gas emissions. The European Green Deal lays the foundations for a sustainable European Union. The growth strategy is intended to show the way to an efficient economy and increase competitiveness. However, the transition to a climate-neutral economy by 2050 poses significant challenges for Europe. At the same time, it presents significant opportunities for innovation and competitiveness. (cf. Alekseev 2024).

2.1.3 European climate targets

Society must innovate in many areas to make clean energy affordable everywhere. This includes advances in nuclear energy, battery storage and CO_2 capture technology (cf. S&P Global 2025). The effects of climate change are no longer subtle; the impact is evident in the form of devastating forest fires, floods and extreme heatwaves (cf. Reuters 2025). More severe consequences are likely in the future (cf. Kaszuba et al. 2024).

Each additional degree of warming introduces a new layer of risk. In order to achieve the climate targets that have been set, scalable solutions that can have a significant impact must be implemented. This entails investing in clean energy, overhauling transportation systems, and enhancing energy efficiency (cf. World Economic Forum 2025). Society must make more informed choices to drive meaningful changes. These changes affect many areas of daily life. They involve rethinking consumption and nutrition (cf. Kerry Health and Nutrition Institute 2025). It could be a sensible goal to identify, measure and model the 100 most effective solutions to combat global warming. Society is currently at a crossroads. Society must decide whether to continue on the current path and accept devastating consequences, or choose a new future that prioritizes sustainability (cf. CISL 2025).

2.2 Sustainability

Sustainability is a powerful term, historically categorized into three main areas: social, ecological, and economic. Yet today, other important dimensions must also be taken into account (cf. Risius 2023). Sustainability is by no means a new concept. The concept of sustainability dates back to the seventeenth century (cf. Dabija & Nastase 2024). The concept of sustainability evolved gradually over many years (cf. United Nations 2023). In 1987, sustainability was identified by the United Nations as one of the most important issues for the industrial and social future.

2.2.1 Concept of sustainability

The following section examines the key definitions of sustainability, the historical development of the term, and the most influential standard works in the field. It evaluates which sustainability definitions are most applicable to industrial contexts. A fundamental understanding of the evolution of sustainability research and the term's various transformations over time provides the foundation for systematizing efforts to achieve sustainability (cf. Gilbert & Perl A. 2018).

According to Mainzer (2021), sustainability is a term with positive connotations; historically, sustainability is understood to mean the preservation of values in the sense of doing business for the benefit of future generations. Against the backdrop of this change, sustainability and IT or AI seem to be at odds with each other. Today's interpretation of sustainability is largely based on the framework set forth in the United Nations' Brundtland Report (UN). Society is in a position to shape development in a sustainable way so that it meets the needs of the present (cf. United Nations 1987).

"*Meeting the needs of the present without compromising the ability of future generations to meet their own needs.*" (United Nations 2023). Sustainability is always related to current business environments. Sustainability means fulfilling today's business needs while simultaneously acting with a future-oriented approach.

2.2.2 Forms of sustainability

The Enquete Report on sustainable and future-oriented development was published back in 1998. It describes concrete measures for shaping sustainable development. It also explained the dimensions that characterize today's image of sustainability and are still highly relevant today (cf. Enquete Commission 2004).

The first component underpinning the concept of sustainability is the environmental dimension. This aspect encompasses the consumption of resources that arise during the production and provision of products or services. From an ecological perspective, the habitat of all people is considered and the diversity of the environment is protected (cf. Stepanek 2022). According to Enste and Potthoff (2021), many opinions agree that man-made depletion of nature has already occurred and that this will have serious consequences for future generations. From an environmental dimension, it is necessary to bring consumption back in line with the limits of the environment.

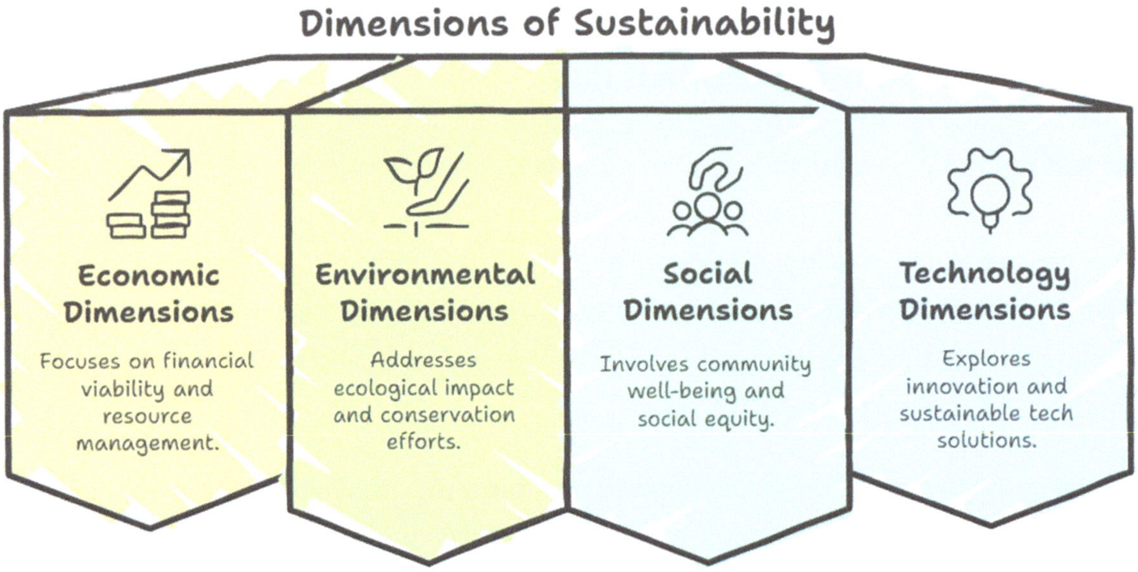

Figure 3: Four pillars of sustainability
Source: Author's own representation.

Though sustainability may seem like an old-fashioned term, it is more relevant than ever in 2025. Sustainability describes the critical examination of the consequences of population growth (cf. Navoco 2024). Technology considered the fourth pillar of sustainability, also encompasses topics related to AI and digitalization.

The social aspect can be understood as the effort to ensure justice across generations over time. The opportunities of future generations should not be restricted by the selfish actions of the current generation. From an economic perspective, sustainability means generating profits in the most environmentally and socially responsible manner, rather than using profits to offset the damage caused by production afterward, since sustainable development must be self-financing. It should not require cross-subsidization from non-sustainable business sectors. From an ecological perspective, sustainability means that the extraction of renewable resources must not exceed the rate of regeneration (cf. Nature sustainalbility 2024). Beyond the classic three-pillar model, other pillars related to sustainability, such as digitalization and technology, can be incorporated. Although the model can be broadened, it remains crucial to address and pursue all dimensions with equal importance.

Besides technology, other dimensions of sustainability exist; however, these are not explored further in this dissertation as they fall outside its focus and scope (cf. United Nations Department of Economic and Social Affairs 2025).

Sustainability is becoming increasingly important in all areas of life. The term sustainability is used in public discussions to refer primarily to ecological aspects of sustainability. The increasing demand for more sustainability through greater climate and environmental protection has led to a series of political initiatives at European level in recent years (cf. Wired 2023). It is anticipated that political and societal demands on companies, particularly regarding environmental and social issues, will continue to grow and become more stringent (cf. Wu et.al. 2024).

2.2.3 Ecological view of sustainability

Around one in two companies is already working intensively on ecological change. Ecological change involves balancing challenges and opportunities equally (cf. KOFA Study 2023). There are no differences by sector or company size and it affects all companies equally (cf. Kasten 2022). Overall, ecological change tends to be perceived as more challenging than digital change (cf. Risius 2023). This may also be because companies' environmental sustainability or behaviors are visible to the public.

Resources that cannot be renewed may only be dismantled if it is ensured that they can be replaced by other resources. This fact must be acknowledged and respected without being exceeded. Resources should only be depleted at a limited rate so that renewal is ensured. The environmental sustainability of companies is reflected in the consumption of resources and the amount of waste they produce (cf. Hauff 2022).

Sustainable companies strive to find solutions and avoid negative impacts in their operations. This could have a negative impact on the environment (cf. Wired 2023). Ecological change is expected to drive increased labor demand in the wind and solar energy sectors. According to Risius (2023), this also affects companies that currently offer neither sustainable products nor sustainable services. The list of possibilities is extensive, ranging from the use of energy from renewable sources to the use of sustainable products and climate-friendly employer services (cf. Reset digital 2024).

2.2.4 Economic view of sustainability

The second view is economic sustainability. This primarily looks at the economy so that it operates within ecological limits in the long term and does not further exploit nature. It is also about a fair economy, which must also be ensured. According to European Commission (2025), a fair distribution of funds and investment in sustainable technologies and innovations are important (cf. Accenture 2024). The creation and development of new jobs and new skills is highly relevant.

Long-term business survival depends on profitability and the ability to generate adequate profits. The performance of companies is reflected in turnover, contribution margins and productivity figures. According to Wired (2023), the ability to deliver also plays an important role. Sustainable companies aim for rapid and lasting business success. Economic opportunities arise from mobility platforms, especially for mobility service providers, who gain greater market penetration and an additional sales channel for their products through their platforms. As a result of improved accessibility, transparency and comparability of individual transport offers, competition in the mobility sector is strengthened in the interests of the end customer (cf. Davis et al. 2024). Other sections of society will benefit from a more efficient transport system, better accessibility and less traffic congestion.

In a globalized world, the need to balance economic growth with environmental protection is increasingly evident. Traditional economic models based on infinite growth are coming under increasing pressure as the environmental impact of human activity becomes more noticeable (cf. De Feo 2019). The economic view of sustainability therefore calls for a fundamental reorientation of economic strategies that considers both the ecological and social dimensions of sustainability (cf. Deutsche Bundesstiftung Umwelt 2024).

The decarbonization of economic sectors, especially logistics, is a central component of this reorientation. McKinnon (2023: 45) argues that the decarbonization of logistics requires a fundamental realignment of supply chains in which economic efficiency is reconciled with environmental sustainability (cf. Zukunftsinstitut 2024). This implies that companies must implement sustainable practices not only for environmental reasons, but also for economic considerations to remain competitive in the long term. The integration of low CO_2 emissions into logistics can not only reduce emissions but also bring cost savings and increased competitiveness (cf. McKinnon 2023).

Another aspect of economic sustainability is the need to integrate ecological limits into economic theory. Gomez-Baggethun and Naredo (2023) emphasize that ecological economics challenges the conventional economic paradigm by placing ecological limits at the forefront of economic analysis. This perspective is crucial as it defines sustainability beyond the mere pursuit of economic growth and emphasizes the intrinsic value of natural ecosystems (cf. Gomez-Baggethun & Naredo 2023). Jackson (2023b) argues that the pursuit of perpetual economic growth on a finite planet is inherently unsustainable and requires a paradigm shift towards ecological balance.

The role of politics in this transformation process is crucial. Grubb and Neuhoff (2023) point out that market dynamics alone are not enough to achieve carbon neutrality. Coordinated policy interventions are essential to steer the transition. This insight highlights the need for policy frameworks that encourage sustainable economic practices and incentivize innovation that benefits both the environment and the economy (cf. McKinnon 2023). In summary, the incorporation of economic sustainability into contemporary economic policies remains a key challenge going forward. Sustainable development requires a holistic approach that reconciles economic growth, social justice and the protection of natural resources.

2.2.5 Social view of sustainability

The third term is social sustainability. Social sustainability is primarily about people's prosperity and how it can be maintained or even increased. The core challenges here are poverty reduction and demographic change. The abolition of immoral forms of work and equal pay. In addition to the economic situation, social factors such as the right to freedom of expression and individual decision-making power are also important. The social system and adequate health insurance are also integral components (cf. European Commission 2025).

Companies are a social structure in which the interests of customers are taken into account. Sustainable companies therefore strive for long-term relationships based on trust. The protection of employees is also an important factor (cf. Wired 2023). In this context, the focus is on ensuring appropriate working conditions and safety-related issues. The employees and their mental and physical manpower ensure the continued existence of the company (cf. Jakob 2023). Companies, including those outside the freight transport sector, send their employees to seminars, professional development courses and certifications, which in turn sustainably increases the quality and thus also the quality of the company. Economic globalization and the rise of digitalization call for a rethinking and restructuring of value chains-development that have already driven steady growth in the transport economics sector in recent years. Companies and their employees are adapting to the changing conditions through professional development (cf. German Society for Quality 2024).

2.2.6 Technology as the fourth dimension of sustainability

Technologies can contribute to ensuring sustainability. Prerequisites must be in place to ensure that technological advancement supports sustainable development. In general, technology is changing society. In addition to communication, working models and work cultures are also changing (cf. Umweltbundesamt 2022). However, increasing networking also increases energy consumption. The transformation of production and infrastructure paves the way for innovative sustainable solutions. At the same time, the increasing demand for technologies and digitalization issues have far-reaching risks. The consumption of natural resources and the energy required for production is increasing. The development of new technology is very complex (cf. International Energy Agency 2025). Analyzing their possibilities and constraints is important for early assessment of major environmental effects. A prerequisite for sustainable development is that digitalization is understood as an environmental policy issue. Politics, business, and research share the responsibility of shaping the future, necessitating coordinated global efforts across all fields to mitigate the adverse impacts of digitalization and meet environmental targets by 2040 (cf. Umweltbundesamt 2022).

In 2025, digitalization will remain a megatrend permeating every aspect of society (cf. Digital Experts 2024). Digitalization also has an impact on sustainability. Megatrends are trends that are changing our society and markets (cf. Diwo 2024), these changes will bring about massive changes (cf. Umweltbundesamt2022). The digitalization megatrend is driving the networking of products and services and the continued increase in internet speed (cf. Thalkur 2025). The changes associated with digitalization are dynamic and contribute to the constant emergence of new and innovative products. Nowadays, the use of technology is indispensable, i.e. it is constantly being further developed and used as a driver for innovation (cf. Umweltbundesamt 2022).

To succeed in the market, companies must ensure long-term profitability while staying at the forefront of technology. They must recognize the opportunities offered by new technologies at an early stage and use them to optimize their performance (cf. Wired 2023).

Technologies such as blockchain and AI play a key role in the sustainable transformation of supply chains (cf. Zielinski 2024: 114). These technologies and innovations are the key to implementing green logistics solutions (cf. Fischer 2021: 78). Sustainable mobility relies heavily on innovations and intelligent transportation systems (cf. Yunex 2025: n.p.). By reducing operating costs, environmentally friendly logistics solutions also help lessen environmental impact (cf. Black 2023). Environmentally friendly supply chain strategies contribute to the long-term competitiveness of companies (cf. Grant 2023: 123).

According to Mainzer (2021), AI has long dominated our lives without many of us being aware of it. Devices such as smartphones and wearables that track health data, store work-related information, and interact with users. There are also examples in logistics that make use of AI, e.g. transport and energy systems with autonomous logistics or robots that explore distant planets. They demonstrate how AI functions shape the field. To examine AI's contribution to sustainability, it is essential to define its specific capabilities in freight transportation

2.3 Mobility

Mobility refers to how people move around. The term is new and has largely replaced the term transportation in general usage (cf. Wilde 2024). Mobility is further defined and distinguished from other topics in the following subchapters.

2.3.1 Concept of mobility

The concept of networked mobility has been added to the digitalization concept. This refers to the organizational and technical connection of means of transport. The concept of mobility is closely linked to changes in means of transport, which in turn are driven by shifts in infrastructure and traffic (cf. Wilde 2024). The problems that give rise to the need for a transformation of mobility include not only the continuing rise in emissions but also the enormous space requirements of HGV traffic, which competes with other usage interests not only in cities, the noise pollution for local residents and the continuing high number of deaths and injuries caused by traffic accidents in Europe (cf. Manderscheid 2024). The concept of mobility is multi-layered and includes many aspects that are particularly relevant to European freight transport. Rammert (2024) explains these critical insights from mobility research and identifies the key dimensions that must be considered for full-scale mobility planning.

According to Wagner and Bechler (2024), mobility concepts of the future will not become self-sufficient, which means that a networked implementation of new technologies in an existing mobility ecosystem is necessary in order to meet social demands. Success factors for a decarbonization of mobility are derived on the basis of mobility concepts (cf. Nickel 2023). Transport companies are encouraged to work on their mobility concepts and, if necessary, adapt them to the changes in the market (cf. Eisenkopf 2023).

The mobility ecosystem is complex, requiring attention to environmental concerns, cost implications, and the challenge of attracting skilled workers. The respective approaches should be stringent and contribute to the optimization of corporate mobility. The Fraunhofer Institute develops measures for long-term corporate success. The mobility concept is based on the individual needs of companies (cf. Fraunhofer 2024).

2.3.2 Forms of mobility

Mobility has become a buzzword today. But mobility is more than that. Many individuals live at a considerable distance from their place of work. They have to satisfy their needs and be mobile. The term *mobility* therefore describes the ability of people to reach their desired destinations. The realization of these possibilities defines transport (cf. Baloise 2024).

Although mobility is one of the key words of our time, hardly anyone would have associated the term mobility with transportation back in the mid-1980s. Today, mobility and transportation are almost synonyms in everyday language. However, both terms convey many different meanings. Mobility extends beyond the realm of transportation, transportation is merely one significant manifestation of it (cf. Gavanas 2024).

2.3.3 Definition of the concept of mobility

Mobility describes the spatial movement of people and objects and has its origins in 18th-century military terminology. *Mobile* in the sense of *movable* or *ready for action*. Mobility is generally defined as physical, psychological or social *mobility* (cf. Liebisch-Gümüs 2024). In the transport context, the term mobility is restricted to the mobility of people outside the home to overcome spatial distances. Mobility also serves to make optimum use of and expand living space (cf. Denteh et al. 2025). In many cases, mobility is also defined as the willingness and ability to change location.

The term mobility is used differently in individual academic disciplines. A distinction can be made between *vertical* and *horizontal* mobility. Mental mobility should also be mentioned as a further form of mobility (Federal Ministry for Climate Action and Environment 2024). The concept of mobility is interdisciplinary and encompasses more than just physical movement. It also includes social, economic and technological aspects that are socially relevant (cf. Reissova et al. 2024).

2.3.4 Mobility in freight transport

Alternative energy sources are increasingly being considered for powering freight vehicles. The aim is to reduce emissions from the transport sector. There are also opportunities for implementation in urban logistics, particularly on the last mile of the supply chain (cf. Adeniran et al. 2023). Reducing CO_2 emissions in the transportation sector is a key contribution to the global climate protection strategy (cf. Jackson 2023a). The implementation of sustainable transport systems requires close cooperation between technology developers and political decision-makers in Europe (cf. International Transport Forum 2025).

Freight transport has increased rapidly, especially since the advent of e-commerce. Road traffic is still the most important mode of transport for the transportation of goods (cf. Mehmedi 2021). The increasing volume of freight traffic has a positive effect on economic power and thus on the quality of life. However, it also has a negative impact on traffic conditions in the city and the environment (cf. Texas A&M Transportation Institute 2023). Road traffic is one of the largest emitters of greenhouse gases (GHG). The need to reduce GHG emissions has become more important than ever. In an effort to reduce greenhouse gas emissions, many EU member states have established zero-emission zones, restricting access for high-emission vehicles (cf. Business.gov.nl 2025). The environmental footprint of freight transport can be significantly improved by using renewable energy (cf. Hofmann 2023). Combining multimodal transport solutions is key to the future of sustainable mobility (cf. McKinnon 2023).

Sustainability in freight transportation requires close cooperation between all actors within supply chains (cf. Grant 2024). Sustainability initiatives in logistics promote the long-term competitiveness of companies (cf. Grant 2023) and can reduce operating costs and increase efficiency (cf. Sampson 2023). Sustainability strategies are becoming increasingly important in the logistics industry (cf. Wensauer 2023). McKinnon (2023) argues that innovative business models and emerging technologies are accelerating the transition to low-carbon freight transport. New technologies and business models are promoting the development of low-carbon logistics. Eco-friendly logistics solutions include the use of low-emission vehicles (cf. Cole 2023). These are essential for reducing emissions in freight transport. Ecological logistics includes the reduction of emissions and the efficient use of resources (cf. Meißner 2024).

Promoting sustainable transportation requires a coordinated effort between policy initiatives and technological advancements. Political measures to promote sustainable transportation solutions are increasing worldwide (cf. Zielinski 2024). Political measures to support environmentally friendly transportation solutions are essential (cf. Weber 2023).

2.3.5 Sustainable mobility

Sustainable mobility is created when attention is paid to autonomous transportation. Last-mile transport, delivering goods from the warehouse to their final destination, is a critical component of the distribution process (cf. Novocargo 2024). The introduction of a combined autonomous vehicle seems advantageous for many reasons. It could reduce vehicle consumption and thus increase efficiency. A reduction in emissions and land use will improve transportation performance for customers (cf. Research and Markets 2024). The development of such technologies is associated with numerous challenges, as the individual characteristics of vehicles have an impact on the entire transportation system. This means that not only economic, but also ecological and social damage can occur if a slow vehicle is used (cf. Deineko 2023). Reducing CO_2 emissions and other greenhouse gases is a major challenge in freight transport. Ensuring the cold chain of foodstuffs in particular poses a major challenge in the latter (cf. Bagwan 2023).

In recent years, strategies have been implemented to reduce CO_2 emissions (cf. Badwan 2023). Mobility is a prerequisite for the functioning of a society and forms the basis of European integration. In addition, the development of sustainable transport infrastructure is essential to ensure the smooth operation of the European Single Market (cf. Laimer and Perathoner 2024). According to the Pro-Rail Alliance, the infrastructure must be expanded. The federal government should electrify 75% of the rail network, including border crossings, by 2030 and implement the 740m network for longer freight trains by 2025 (cf. Railfreight.com 2024).

2.3.6 Mobility competition in Europe

Research is currently being conducted into topics such as electric mobility and alternative drive systems. The focus is on the further development of electric vehicles and other environmentally friendly drive technologies such as hydrogen and fuel cell technologies (cf. Fraunhofer IFAM 2024). Intelligent traffic management systems are being developed, which are also used for traffic monitoring and optimization. Nell (2024) emphasizes the need for sustainable transportation concepts that align with set requirements and environmental objectives. This change will have an enormous impact on the design of mobility services and vehicle drive technology. The manifestation of these demands in society's self-image will have an impact on the increased expansion of public transport (cf. Gutzner and Todsen 2021).

As stated by Kille (2024), transport infrastructure is key to the economic development of individual European nations. The infrastructure includes roads, railways, waterways and airports. In Europe, the highway network forms the backbone of the road infrastructure. It enables goods to be transported quickly and efficiently over long distances. The dense rail network ensures efficient connections to cities. The Rhine and other inland waterways serve as vital routes for freight transport. Austria, as a landlocked country bordering the Alps, has a well-developed road and rail infrastructure (cf. Ungerboeck 2023).

Alpine passes and tunnel connections are of central importance for transportation. The well-developed railroad lines are important transit routes through the Alps. Switzerland, another landlocked country bordering the Alps, has developed a similar transport infrastructure. Switzerland combines a dense road infrastructure with one of the world's most efficient and compact rail networks (cf. Kille 2024).

2.4 Freight transport

2.4.1 Definition of freight transport

There are many definitions of the term freight transport. The valid definition for this work: "Freight transport comprises the transportation of goods of all kinds between different locations by road, rail, water, air or pipeline" (Federal Statistical Office 2024). According to this definition, the most important task of freight transport is to ensure the transportation of goods from production to the customer. With the steady increase in digitalization and the associated networking, the challenges of freight transport in terms of efficient material flow and freight transport have increased (cf. BVL 2024).

2.4.2 Forms of freight transport

Freight is transported by various modes of transport. According to Muchna et al. (2021), the individual modes of transport can be described.

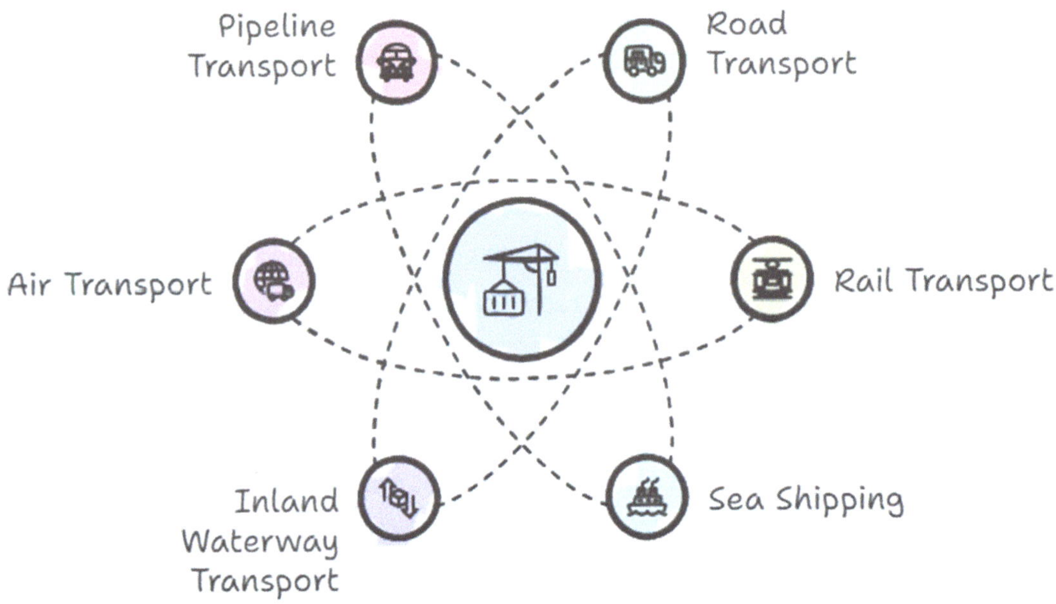

Figure 4: Forms of freight transport
Source: Author's own representation.

- **Road** haulage**:** Road haulage covers all road haulage in distribution and long-distance transport and is also referred to as road transport.
- **Rail transport**: Rail employees are the main mode of long-distance rail freight transport. Intercontinental rail transportation between China and Europe has also been established.
- **Maritime shipping:** Large volumes of goods are transported by maritime shipping exclusively on long-distance, mostly intercontinental sea routes.
- **Inland navigation**: Inland navigation transports goods by waterway in local and long-distance traffic within the European continent.
- **Air transport:** Air transport is used to transport goods, primarily in international long-distance transport by aircraft.
- **Pipework** transport**:** Pipework transport is realized with the help of stationary pipework, which is used exclusively to move the transported goods.

A mode of transport is understood to be the entirety of the undertakings that make use of a specific infrastructure. Transport infrastructure refers to all fixed installations and facilities that serve to carry out transportation. It includes all transport routes and stations such as their entries, departures and transshipment terminals. Vehicles that, together with transport infrastructure, enable traffic to function are considered means of transport. They are primarily used to pick up the goods to be transported (cf. Jusda 2024).

Alongside conventional monomodal freight transport, combined transport is another important method. This refers to the integration of two or more means of transport and, as a rule, also modes of transport. The forms of combined transport include large containers, piggyback transport and bimodal techniques (cf. Hupac Intermodal 2021). Container traffic in the hinterland of European seaports is particularly relevant during the EU's eastward expansion (cf. Izotov 2024).

2.4.3 Differentiation between freight transport and logistics

The term *freight transport* must be distinguished from the term *logistics*. Göpfert (2024) notes that 'logistics' became an established term in the 1970s, later transitioning to supply chain management (SCM) in the 1990s (cf. Adams & Ellis 2019). While the academic debate on logistics began noticeably in the 1970s in German-speaking countries, it was not until two decades later in the 1990s that supply chain management dominated the discussion (cf. Gattorna 2019). The terms *logistics* and *supply chain management* have numerous definitions and areas of application (cf. Kummer 2023). However, they all pursue the same goal, namely, to optimize the transport of goods to save costs and minimize emissions. While sustainable transport strategies promote environmental balance and competitiveness (cf. Schneider 2024), they must also be adapted to local conditions (cf. Curtis 2024). These transportation strategies include both technological and political measures (cf. Ryley 2023).

Freight transport is part of transport logistics, which, according to Göpfert (2024), is a modern management concept for organizing the flow of goods, information and finance. Logistics can therefore also be seen as the backbone of value chains (cf. Hensher & Button 2023).

2.4.4 The state of green energy in freight transport

2023 was not only marked by visible consequences of climate change but also recorded as the hottest year in history. 2023 was also the year with the highest measured CO_2 emissions worldwide. Nevertheless, many efforts and immediate action are needed to counteract this (cf. Steininger 2024). Sustainability and green logistics are on everyone's lips and have become an integral part of today's language. Green logistics refers to the comprehensive transformation of strategies and systems to establish environmentally friendly logistics processes. By incorporating sustainable practices into every stage, this transformation expands logistics to encompass its vital ecological dimension. (cf. Seeburger Logistics 2024). Collaborative planning and participation are crucial for the implementation of sustainable transportation strategies and are crucial for the implementation of sustainable mobility solutions (cf. Tumlin 2024).

Daily news reports show that Deutsche Bahn has set up its first hydrogen filling station for locomotives in Tübingen. The path has thus been set for increasingly climate-friendly rail transport; however, alternative technologies also exist (cf. Tageschau 2024). If overhead lines were installed across all rail routes, the challenge would be to maximize the input of electricity from renewable sources. Green supply chains make a significant contribution to reducing the environmental footprint of companies (cf. Grant 2024). Green supply chain strategies increasingly include the minimization of waste and emissions (cf. Sampson 2023). However, only 60 percent of the German rail network is electrified, and even less in Europe. Diesel locomotives have been used for the remaining 40%. And the combustion of diesel fuel produces climate-damaging carbon dioxide.

Environmental organizations such as Greenpeace welcome the efforts of the European rail companies. Even today, rail is already the most climate-friendly means of transportation. However, too little has been invested in the transformation of the railways over the decades. Compared to other countries, Germany is lagging in the expansion of overhead lines. According to Greenpeace, improving the traction power network should be a top priority. The aim should be to electrify 79% of the lines. 600 additional kilometers per year are needed (cf. Greenpeace 2024). These green strategies contribute greatly to reducing environmental pollution and cutting costs (cf. Schutze 2023). Green logistics strategies also include minimizing empty runs, optimizing transport routes and using environmentally friendly means of transport (cf. Cullinane 2024).

For Greenpeace, battery-powered trains are the better choice than hydrogen-powered locomotives. Greenpeace reports that hydrogen trains cost 80% more than battery trains over 30 years and suffer from 60% energy loss during the electricity-to-hydrogen conversion process, making them less efficient (Greenpeace 2024). In addition, hydrogen is still in very short supply for the foreseeable future and is urgently needed elsewhere, for example in industry or to power large ships (cf. Tagesschau from July 6, 2024a).

During the transition period, Deutsche Bahn aims to use biodiesel made from agricultural and food production waste to power diesel locomotives. Greenpeace views this with skepticism. "Waste" also includes materials such as waste wood or straw, which could be put to better use elsewhere (cf. Tagesschau from 30.05.2022). According to Dr. Simon Experte 7 from hySolutions (2024), there are currently over 15,000 diesel locomotives running in Europe, which are to be replaced by climate-neutral rail vehicles by 2050. Load profiles of trains need to be examined more closely to accelerate technological development.

Siemens Mobility is testing the use of hydrogen for rail transportation. A newly developed train combined with a mobile filling station is currently being tested. The hydrogen train will run on a trial basis in the Tübingen area from August 2024 (cv. Energy news 2024). It is expected to save around 330 tons of CO_2 (cf. Deutsche Bahn 2024a). The rail company wants to use hydrogen trains to replace its diesel locomotives and significantly reduce CO_2 emissions on the railways. The German Federal Ministry for Digital and Transport (BMVI) has promised to fund the project (cf. Energate 2024). The Siemens Mobility train is based on the Mireo Plus regional multiple units. The maximum speed is 160 km/h. Deutsche Bahn states in its press release that the hydrogen drive will reduce CO2 emissions by around 330 tons per year (cf. Deutsche Bahn 2024b). Innovative technologies and processes are crucial for the implementation of sustainable logistics solutions (cf. Wensauer 2023).

Due to the modular design, power sources can also be combined in the vehicle. A study by the German Electrical Engineering Association showed that fuel cells are not always the best alternative to fossil fuels (cf. Handelsblatt 2024). Green transportation technologies are the key to achieving sustainable mobility (cf. Zielinksi 2024). Given that 2023 was both the hottest year on record and the year with the highest global CO2 emissions, alongside particularly visible climate change impacts, far more ambitious and immediate action is essential. (cf. Steininger 2024).

2.5 Interim conclusion

Green energy refers to energy sources that come from natural, renewable resources such as water, wind and sunlight and do not release CO_2 emissions. Green energy differs from renewable energy, which can also come from renewable sources but is not necessarily emission-free. Biomass, for example, is considered renewable but causes emissions. As part of the Green Deal, the EU is pursuing the goal of reducing greenhouse gas emissions by 55% by 2030 and achieving climate neutrality by 2050. These targets are part of international agreements, such as the Paris Climate Agreement, and require extensive investment in innovation and technologies. The European Green Deal represents a comprehensive growth strategy for the EU that not only aims to achieve climate neutrality but also to strengthen economic competitiveness.

A distinction is made between green, clean and renewable energy. Green energy is derived from natural, emission-free sources. Clean energy, on the other hand, includes all forms of energy that cause little or no emissions, regardless of their renewability. Renewable energy sources are replenished naturally, but are not always clean or green. For example, hydropower plants can be ecologically problematic even though they are considered renewable. Technological and societal changes are needed to achieve climate targets. Climate change is a major challenge, but it requires innovation in the areas of clean energy, mobility and digitalization. To keep global warming below 2°C following the Paris Agreement, investments must be made in green technologies and sustainable transportation systems.

Sustainability is classically divided into three dimensions: ecological, economic and social. Ecological sustainability means that the consumption of natural resources must not exceed the rate of regeneration. Economic sustainability requires that economic activities take place within ecological limits and that social justice is guaranteed. Social sustainability aims to secure long-term prosperity and create fair living conditions. Recently, technological sustainability has also been discussed as the fourth pillar, which emphasizes the use of innovative technologies to promote sustainable development. Mobility encompasses the movement of people and goods, taking environmental and energy aspects into account. In freight transport, alternative energy sources and low-emission drive systems are key to reducing greenhouse gas emissions. Concepts such as the Cargo Tube and Hyperloop offer environmentally friendly solutions.

Freight transport plays a central role in transport logistics and is essential for value chains, as it ensures the flow of materials from the producer to the consumer. Digitalization and networking have increased the demands placed on freight transport, particularly regarding the efficiency of material flows and the reduction of emissions. With the development of the concept of logistics into supply chain management in the 1990s, a modern management concept was established that forms the basis of today's transport logistics.

3 Freight transport in Europe

The terms Europe, European Union (EU) and European Single Market must be clearly distinguished from one another. While Europe is defined as a geographical continent, the EU refers to a political and economic union of currently 27 member states. The European single market comprises all countries that participate in the free movement of goods, services, capital and people within the EU and in associated countries such as Norway, Iceland and Liechtenstein. This analysis takes into account the entire European region and is therefore not limited to the European Union as a political entity. This differentiated perspective enables a more comprehensive analysis that also includes non-EU European states (cf. Wozniak-Jechorek 2024).

Europe, with an area of approximately ten million square kilometers, is the world's second smallest continent and is bordered by the Atlantic Ocean to the west and north (cf. EIZ Rostock 2024). European freight transport is still dominated by road transport, while a significant shift to rail transport has not yet been achieved.

The importance of alternative modes of transport varies greatly between European countries. In the Baltic countries, more than 50% of freight transport is handled by rail and inland navigation, which is due to a correspondingly developed infrastructure. By contrast, in larger countries such as France and Spain, these modes of transport account for less than 13% of the total share (cf. Federal Statistical Office 2024). The integration of ecological, economic and social aspects into transport planning is essential for the sustainable development of European freight transport (cf. Jackson 2023a). Addressing this issue requires both greater investment in rail and waterborne infrastructure and the implementation of policy measures aimed at promoting environmentally sustainable modes of transport.

The European Union constitutes a treaty-based association of sovereign states, with its foundations dating back to May 9, 1950. It acquired its official designation upon the entry into force of the Maastricht Treaty in November 1993. The EU consists of 27 member states with a total population of over half a billion people (cf. European Union 2024).

The EU is facing significant environmental challenges in the face of climate change. In 2023, 31% of global greenhouse gas emissions were attributable to electricity and heat production, followed by freight transport with a share of 19% across all modes of transport (cf. Our World In Data 2023). To overcome these challenges, the European Green Deal was presented in December 2021, which aims to achieve climate neutrality by 2050. Achieving this goal necessitates the implementation of more ambitious emission targets that take into account the diverse baseline conditions across individual member states (cf. UBA Dessau 2023). Despite these efforts, the targeted shift from traffic to rail has been slow to date. In 2023, the share of rail freight transport in the EU was only 17 %, while the share of inland navigation fell from 7 % to 6 %. At the same time, the share of road transport increased from 74% to 77% in the same period, which illustrates the challenges of implementing sustainable transport policy (cf. Federal Statistical Office 2024).

The promotion of rail transport as a low-emission mode of transport remains a central measure of EU environmental policy. To realize the climate ambitions of the Green Deal, comprehensive investments in infrastructure combined with more far-reaching political action are indispensable (cf. Varnavskii 2024).

The European Single Market is the largest common economic area in the world in terms of gross domestic product (GDP). It enables the free exchange of goods, services, capital and people between member states. Since the founding of the EU, 22 other countries have joined the Single Market, creating a dynamic and competitive economic area (cf. Tovias 2024). The ongoing integration and refinement of these fundamental freedoms enable the Single Market to play a pivotal role in safeguarding Europe's economic stability and strengthening its position in the global economy. The illustration depicts the four freedoms of the European Union's (EU) single market.

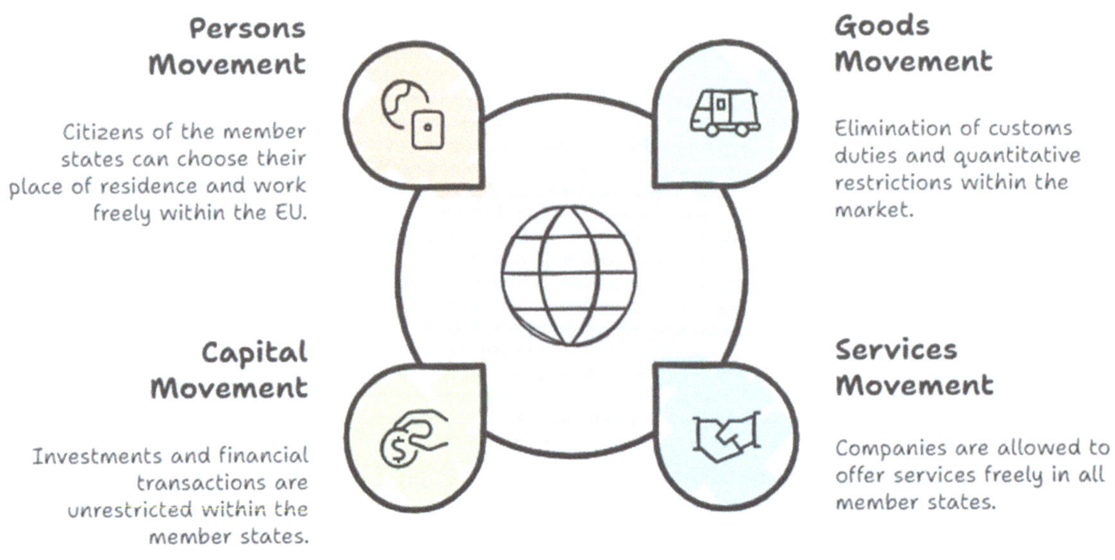

Figure 5: EU Single Market
Source: Author's own representation.

This is primarily grounded in the four fundamental freedoms, which form the cornerstone of the Single Market. Anything that restricts the free movement of fundamental freedoms is seen as an obstacle to the Single Market. A major impediment to the Single

Market lies in the divergent interpretations and the sluggish flow of information concerning its regulations across the individual member states. To counteract this, Member States and the Commission are working together to raise awareness and improve their information channels (cf. Tovias 2024). The effort of the EU can generally be divided into four areas of rights.

Freedom of movement for people, which allows citizens of EU member states to choose their place of residence and work freely within the EU. Freedom of movement for goods, ensuring the elimination of customs duties and quantitative restrictions within the internal market. Freedom to provide services, allowing companies to offer their services freely across all member states. Freedom of capital movement, enabling unrestricted investments and financial transactions within the EU. At the center of the illustration, a global symbol emphasizes the significance of these freedoms as the core pillars of integration and cooperation within the EU (cf. Kille 2024).

3.1 Mobility in the transportation sector

The comparative analysis of all market segments reveals a structure that closely mirrors the patterns identified in passenger transport. In contrast to passenger transport, however, transport performance is measured in a different unit. This metric represents tonnage-distance, obtained by multiplying the weight of the goods transported (in tons) by the transport distance (in kilometers) (cf. Fedoskin et al. 2024):

Transport Performance (tkm) = Transported Goods (t) × Distance Traveled (km)

By applying this formula and evaluating the transport performance in metric ton kilometers (tkm), the individual modes of transport can be better compared in terms of their use and relevance for freight transport (cf. Umweltbundesamt 2022a). Table 1 lists all modes of transport used to transport goods within Europe in the last three years. Maritime shipping was deliberately not included, as this is not primarily a national transport mode that could be substituted by other modes of transport and thus compared with them. The table shows that freight transport by air has only a very small and negligible share (< 0.5 %) of the total transport volume (cf. Xiong et al. 2024.).

This is not a temporary low point in the last three years, but the share has remained consistently low for more than a decade. The second-lowest traffic type can also be disregarded, as it only accounts for < 3% of the total traffic volume. Furthermore, long-distance pipelines are used exclusively for the transportation of one type of goods, crude oil.

Year	Road traffic	Rail traffic	Barge	Pipelines	Air traffic	Total
2019	498,6 tkm (71,1%)	132,8 tkm (18,9%)	50,9 tkm (7,3%)	17,6 tkm (2,5%)	1,6 tkm (0,2%)	701,5 tkm
2020	487,4 tkm (71,7%)	128,0 tkm (18,8%)	46,3 tkm (6,8%)	16,7 tkm (2,4%)	1,5 tkm (0,3%)	679,9 tkm
2021	505,7 tkm (70,7%)	141,0 tkm (19,8%)	48,2 tkm (6,7%)	15,7 tkm (2,2%)	1,8 tkm (0,6%)	712,4 tkm
2022	503,1 tkm (84,7%)	144,3 tkm (20,3%)	44,1 tkm (6,2%)	17,7 tkm (2.4%)	1,6 tkm (2,4%)	710,8 tkm
2023	480,7 tkm (71,2%)	134,4 tkm (19,9%)	41,5 tkm (6,1%)	17,3 tkm (2,5%)	1,5 tkm (0,3%)	675,4 tkm
2024	480,7 tkm (71,2%)	134,4 tkm (19,9%)	41,5 tkm (6,1%)	17,3 tkm (2,5%)	1,5 tkm (0,3%)	675,4 tkm
2025	490,0 tkm (71,7%)	136,0 tkm (19,9%)	39,5 tkm (5,8%)	17,0 tkm (2,4%)	1,6 tkm (0,2%)	684,1 tkm

Table 1: Transport performance within Germany
Source: Based on the Federal Statistical Office, 2025.

The amounts are shown in billion ton-kilometers. It can be seen that freight transport by air has only a very small and negligible share (< 0.5 %) of the total transport volume. The values for 2025 are extrapolated based on data from recent years.

The Swedish freight transport sector is in transition to vehicles without fossil fuels. The introduction of electrified vehicles is currently seen as an important contribution to achieving this goal. However, the necessary transition is associated with great uncertainty for many stakeholders, which is largely due to knowledge gaps (cf. Stelling & Brunner 2023). Transitioning to a transportation system that operates without reliance on fossil fuels constitutes a critical step toward climate neutrality. The Swedish government has set itself the goal of reducing GHG emissions from domestic transport by 70 percent by 2030 compared to 2010 (cf. Regeingen 2017). Between 2010 and 2019, emissions from heavy goods vehicles fell by 31.1 percent (cf. Fit for 55 2022). The introduction of electric trucks and other low-emission vehicles is driving sustainable mobility forward (cf. Hall 2023).

The electrification of freight transport has only just begun, but is seen as an important contribution to further reducing carbon emissions (cf. Stelling & Brunner 2023). Henderson and Clark (1990) highlight that change is more difficult to implement when existing assets are rendered obsolete, necessitating the acquisition and integration of novel resources (cf. World of Work 2024). In a changing market, development is continuous, unpredictable and rapid. It can therefore be a challenge for knowledge development to keep pace. At the same time, conditions change, making the knowledge gained less useful. Hence, it is essential to analyze the current knowledge and learning conditions of stakeholders engaged in the transition of the Swedish freight transport sector, so as to propose effective measures for their advancement (cf. European Parliament 2023).

Ensuring timely delivery in response to transport demands constitutes a strategic imperative; however, the provision of additional capacity invariably incurs significant cost implications. One challenge is to optimize the flow of goods while avoiding changes in capacity (cf. Ebert & Wu 2023). If logistics is understood as a process of controlling the flow of materials in the value chain, the flow of information is also of fundamental importance. Without the integration of contemporary information and communication technologies, the effective implementation of global value chains would not be feasible (cf. OECD 2025).

3.1.1 Energy consumption and emissions

The term *energy consumption* is used several times in this dissertation, as it is primarily used to make a comparison across different modes of transport. The amount of energy is measured in *joules*. It is possible to convert this unit into other units. The following unit conversions can serve as examples (cf. Quaschning 2018):

- 1 kilowatt hour (kWh) = 36 megajoules (Mj)
- 1 hard coal unit (HCU) = 293 Mj = 8.14 kWh

The term hard coal unit describes the energy released when one kilogram of hard coal is burned (cf. Bockhorst 2022). It should be noted that there are differences between primary energy and final energy. Primary energy refers to the total energy input needed across the entire supply chain to deliver usable final energy. Among these processes are the extraction, refinement, and transport of energy carriers like oil, as well as the generation and transmission of electricity. This also requires energy. A certain amount of energy may be unintentionally lost until it is collected and used at the destination. The term 'final energy' designates the energy that reaches the consumer in a usable form, following its transformation and distribution (cf. Grimm 2017).

The figure shown below describes the sub-sectors of the transportation industry that are broken down by different energy sources. Final energy consumption of transport in Germany by mode of transport (in petajoules).

Year	Road traffic	Rail traffic	Barge	Pipelines	Air traffic	Total
2019	2.224 PJ 77,4%	52 PJ 1,8%	11PJ 0,1%	120 PJ 4,2%	436 PJ 16,5	2.873 PJ
2020	1.998 PJ 84,8%	47 PJ 1,9%	10 PJ 0,1%	115 PJ 4,8%	187 PJ 8,4%	2.357 PJ
2021	2.046 PJ 83,8%	48 PJ 1,9%	10 PJ 0,1%	118 PJ 4,8%	220 PJ 9,4%	2.442 PJ
2022	2.085 PJ 81,6%	49 PJ 1,9%	10 PJ 0,1%	122 PJ 4,7%	290 PJ 11,6%	2.556 PJ
2023	2.100 PJ 80,9%	50 PJ 1,9%	10 PJ 0,1%	125 PJ 4,8%	310 PJ 12,3%	2.595 PJ
2024	2.150 PJ 78,4%	51 PJ 1,8%	10 PJ 0,1%	128 PJ 4,6%	330 PJ 15,1%	2.742 PJ

Table 2: Energy consumption by mode of transport
Source: Based on the Bundesamt für Digitales.

The percentage and absolute share of fossil fuels used to generate energy has decreased over the last few decades and 78% of total energy consumption is covered directly by electricity. In road transportation, the use of electricity has continued to increase over the last three years. Renewable energies play a rather subordinate role (cf. DVZ 2023). Over 90% of all means of transport continue to generate energy by burning fuels based on mineral oil. This mechanism suffers from low efficiency, as it results in the generation of waste heat that cannot be harnessed and must be expelled. Less than 50 % of the total energy generated can be used to operate the corresponding means of transport (cf. Umweltbundesamt2022c).

At 19.4%, CO_2 emissions from freight transport are in third place. The energy industry tops the list with 32.2 %. Industry is in second place with 24.2 % (cf. Haße 2022). Here, it is crucial to highlight that the distribution of shares across sectors adheres to the polluter pays principle. This means that the electricity required for train operation is first produced in power plants and then used by the rail company. For this reason, the carbon dioxide generated is allocated to the energy sector and not to the transport sector. Only diesel trains are accounted for directly in the transport sector, given that their engines produce energy through the direct combustion of fuel (cf. Umweltbundesamt2022c).

3.1.2 Infrastructure expansion

A fully operational and intact rail network serves as the cornerstone for the seamless functioning of rail transport. High-performance steel, which is very heavy, is used for a track. One kilometer of laid track weighs around 120 tons. The production of this steel generates around 240 tons of CO_2 (cf. Güterbahnen 2023). When rails are laid, the subsoil is prepared accordingly. This assessment has yet to factor in the expenditures linked to construction activities involving the use of large-scale, fossil fuel-dependent machinery. From this, it can be concluded that the emissions emitted amounted to several million tons of CO_2.

3.1.3 Cargo tubes

There are various approaches to counteracting GHGs, e.g. the use of hyper-loop and cargo tube systems. The cargo tube is an innovative approach for reducing CO_2 emissions and for efficient and safe transportation. A cargo tube is a hyperloop-based transport system designed for the transportation of goods (cf. Duin 2023). Hyperloop is a transport system that can be used as a supplement to the transport network for the fast transportation of goods over short distances. An Hyperloop system consists of three main elements: Tubes, gondolas and terminals (cf. Motwani 2021). According to Motwani and Gupta (2021), the tube is a sealed low-pressure system. The gondola is a container that is under atmospheric pressure and travels within this tube largely free of drag. Magnetic levitation minimizes mechanical friction. A terminal handles the arrivals and departures of the gondolas (cf. Duin 2023).

The cargo tube is based on the Hyperloop concept and has the potential to reduce the overall energy requirements of transport and thus minimize greenhouse gas emissions (cf. Raza 2025). However, a cargo tube relies on conventional rail technology to guide the vehicle in the tube. By leveraging existing technologies, cargo tubes can substantially reduce development costs while facilitating the rapid expansion of transport networks. The concept of the cargo tube leads to a balance of performance, efficiency and durability to provide a freight transportation system for industrial logistics (cf. HyperloopTT 2025). Numerous questions must be addressed in the planning of the infrastructure for a cargo tube system, encompassing aspects from route alignment to capacity planning and operating costs (cf. Duin 2023).

According to Fleischmann & Klose (2021), cargo tubes offer an opportunity to reduce transport costs through optimized routing and minimal energy losses. The underlying technical infrastructure of cargo tubes enables a high level of reliability and safety in freight transportation. By using cargo tubes, CO_2 emissions can be reduced by up to 90% compared to conventional truck transportation. Cargo tubes also offer an effective solution for reducing CO_2 emissions in freight transport (cf. McKinnon 2023). The use of renewable energy for cargo tubes contributes significantly to sustainability (cf. McKinnon 2023). They offer considerable benefits by easing traffic bottlenecks and decreasing pollution in metropolitan areas. However, cargo tubes require specific geographical and infrastructural planning to be optimally integrated (cf. Rodrigue 2021). They also offer a modern solution for optimizing supply chains through faster and more reliable transport methods (cf. Coyle 2021).

As highlighted by Macharis (cf. 2021: 67), the ecological advantages of cargo tubes are reflected in marked decreases in air pollutants and improved air quality (cf. Hensher & Mulley 2022). According to Randolph (cf. 2022), cargo tubes offer advanced technology for improving the efficiency of freight transport. Consequently, cargo tubes contribute to lowering environmental impacts (cf. Rodrigue 2022) and mark a crucial development in the evolution of logistics and supply chain practices (cf. Behrends 2022).

Their implementation can reduce traffic congestion and emissions in cities (cf. Banister 2022). They therefore offer an economically and ecologically efficient solution for freight transportation (cf. Blauwens 2022). By virtue of their environmentally conscious design, cargo tubes play a role in decreasing CO_2 emissions (cf. Monios 2023). They offer an advanced solution for modern freight transport (cf. Savy 2023) and also contribute significantly to a good image. Cargo tubes use advanced technologies to increase efficiency and sustainability in intermodal transportation (cf. Crainic 2023). Through the reduction of environmental burdens, cargo tubes play a crucial role in supporting the achievement of international climate goals (cf. HyperloopTT 2025).

3.2 Freight transport in Germany

The procedure for comparing all market segments is the same as in the previous chapter. In contrast to passenger transport, however, the transport performance is measured in a different unit (cf. Umweltbundesamt 2024). This unit is made up of the transported goods weight in tons multiplied by the distance covered in km. The first type of transport to be considered, with a share > 5 %, is shipping on inland waterways within Germany. It accounts for a share of approx. 7 % over the last three years. The transport performance of inland navigation has been falling continuously since 2011 (cf. Federal Statistical Office 2022).

Rail transport in Germany accounts for only 18% of the modal share, with the trend continuing to decline. As Fockenbrock wrote in November 2020, the rail network is a construction site (cf. Fockenbrock 2020). It is common for such construction sites to operate for multiple months. This leads to interruptions on important route corridors (cf. CM Logistics 2024). These disruptions, coupled with associated delivery deficiencies, have led more companies to relocate their shipments to road transport. The share of total transport volume has remained constant at around 18-19% for many years (cf.

BMDV 2023). The largest proportion of freight transport in Germany is carried out by road.

This illustration enhances the understanding of how freight transport is distributed and prioritized across different modes. If the pure transport performance is considered first, the impression is that most goods are transported over longer distances (> 150 km), as long-distance transport accounts for the largest share (64.7%) of transport performance. Most goods are transported by local transport (cf. BMDV 2023).

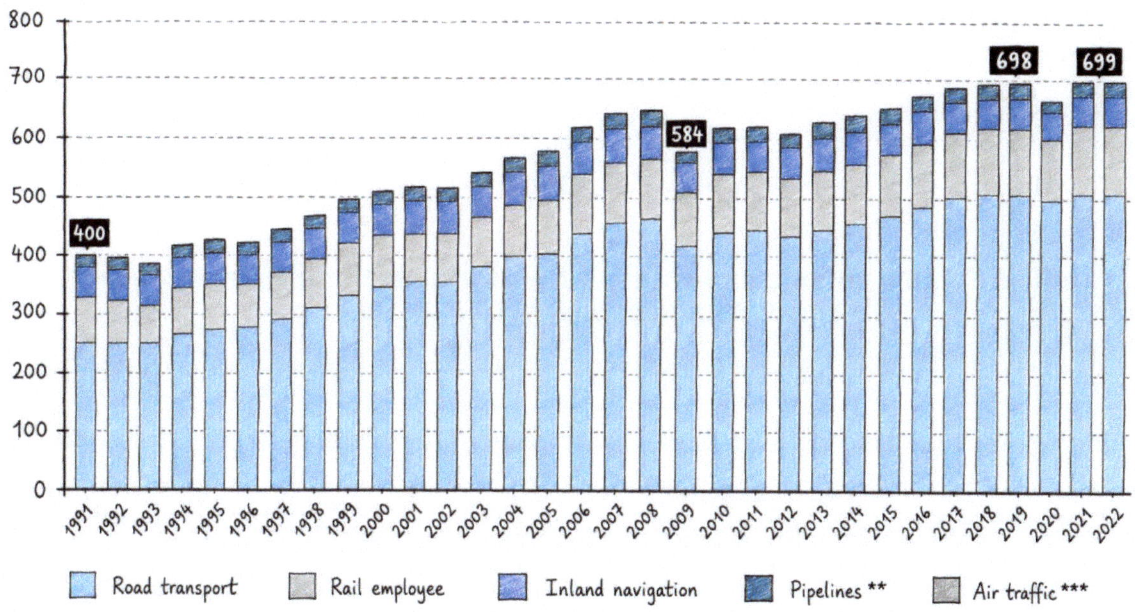

Figure 6: Freight transport performance by mode of transport in Germany
Source: Author's own illustration based on BMVI.

Road transport is also the dominant mode of transport in Germany with the largest transport volume of over 70%. This highlights that in the context of long-distance transport, distance plays a dominant role, as transport performance, and thereby the associated load, remains high even when freight volumes are relatively low (cf. Szaruga 2024).

The figure presents a time series illustrating the modal split of transport from 1991 to 2022. It includes data on various modes of transportation, such as road transport, rail transport, inland waterways, pipelines, and air traffic. It is evident that road transport constitutes the largest share, followed by other categories such as rail transport and inland waterways. Over the years, a consistent increase in the total transport volume has been observed, rising significantly from 400 in 1993 to 699 in 2022. Air traffic accounts for only a small but steadily growing proportion.

One of the main objectives of European transport policy is to reduce GHG emissions by shifting 30% of transport over distances of 300 km or more from road to more environmentally friendly modes of transport (cf. BMDV 2023: n.p.).

Road transport	**Volume (tkm and t in comparison)**
Local transport (up to 50 km)	33.2 billion tkm (11.7%) 1703.7 billion tons (55.0%)
Regional transport (from 51 to 150 km)	67.4 billion tkm (23.6%) 756.9 billion tons (24.5%)
Long-distance traffic (over 150 km)	184.5 billion tkm (64.7%) 634.5 billion tons (20.5%)
Total (tkm) Total (t)	285.1 billion tkm (100%) 3095.1 billion tons (100%)

Table 3: Traffic volume and transport performance in freight transport
Source: Based on: BMDV 2023: 241-247.

Nevertheless, road transport continues to dominate long-distance freight transport in Germany (cf. Banister 2023). Strategies for shifting more freight to rail have prompted significant discourse within the transport economics community (cf. Ebert & Wu 2023). In Germany alone, almost 47,000 companies are involved in commercial road haulage. More than 720,000 vehicles offer loading capacity for more than 10.3 million tons of goods (cf. Stölzle, Schöder, Häberle 2024). According to the Pro-Rail Alliance (2024), reliable framework conditions have been created for European freight transport. The federal government should now advocate uniform framework conditions for combined transport at EU level. This also includes truck dimensions and weights.

The path toward a sustainable transportation system depends on reducing reliance on fossil fuels in favor of renewable energy alternatives (cf. REN21 2024). Adapting to climate change is one of the biggest challenges for freight transport (cf. Ryley 2023). Adapting to climate change is a key challenge for the entire transportation industry (cf. Carter 2023). The adaptation of transportation infrastructure to climate change is a growing trend (cf. Ryley 2023). Depending on the length of the overhead line network, catenary trucks can achieve economic advantages over diesel trucks, especially over long distances. Greening the supply chain requires a holistic view of all transportation processes (cf. Sampson 2023).

3.3 Freight transport in Europe

The reliable movement of goods within Europe depends fundamentally on the functionality and resilience of freight transport systems. Freight transport is an important economic sector for the European economy and comprises around 1.2 million companies with around 10.5 million employees (cf. Federal Statistical Office 2024).

Freight transport leads to severe environmental pollution (cf. Brunner et al. 2020). This is why innovative traffic models and strategies are needed that focus on mobility. The objective of European transport policy is to achieve sustainability and efficiency through the seamless integration of all transport modes (cf. Wozniak-Jechorek 2024). Mobility and digitalization are just some of the issues addressed by European transport policy (cf. Dorfmann 2022). Climate-neutral trains also play a major role in reducing CO_2 emissions. One of the central challenges is to assess the suitability of various *green* propulsion technologies (cf. hysolution 2024). With a 72% share of the transport sector, heavy good vehicle (HGV) traffic is in first place despite a Sunday driving ban (cf. Gomes et al. 2023). The principal argument advanced by freight operating companies concerns the low wage levels paid to drivers. Although a truck emits 110x more CO_2, it still defends its first place in the transportation ranking (cf. CM Logistics Group 2024).

Two years after the launch of the first hydrogen-powered railcars, Deutsche Bahn is also trying out this new technology. Together with Siemens, Deutsche Bahn will be operating a number of test trains in 2024 (cf. Fockenbrock 2020). Hydrogen-powered railcars have the potential to replace diesel trains in the future, thereby contributing to a reduction in the carbon footprint. The special feature of the Siemens train is that it can run on batteries as well as electricity from the overhead line. According to Fockenbrock (cf. 2020.), the French company Alstom is a pioneer in the field of fuel cell-powered trains. From today's perspective, fuel cell trucks will only be able to operate competitively in the medium term beyond niche applications with continued government intervention (cf. IEA 2025). The economic feasibility of using electrolytic hydrogen produced domestically is not anticipated until 2030. The availability of cheap imported hydrogen from renewable energies must be considered questionable in the medium term in view of strong competition for use (cf. Jöhrens 2022). The Austrian Federal Railways (ÖBB) successfully completed a ten-week test run of the hydrogen train (cf. Fockenbrock 2020). Except for pipeline transport, volume increases are anticipated across nearly all modes of transport over the next two years (cf. Railway Technology 2020).

The proportion of goods transported by rail in Europe is less than 20 percent. In addition, wagons are still coupled manually by hand (cf. CM Logistics Group 2024). This is very laborious and time-consuming. This phenomenon notably impedes cross-border traffic in particular (cf. Gatta et al. 2019). Rail freight transport is mainly profitable over long distances (cf. Witte 2024). Cross-border rail freight transport has other challenges that need to be addressed. Train crew and locomotives are often exchanged upon crossing European national boundaries. Different standards and requirements apply to train drivers in many countries. These requirements do not exist for truck drivers. This contributes to the deceleration of cross-border rail freight, while also increasing its complexity and associated costs (cf. Pro-Rail Alliance 2024).

The rail network must also be further expanded so that cross-border rail transport can continue to grow. Projects such as the expansion of the Brenner highway must be implemented quickly (cf. Euronews 2025). The electrification of border crossings must also be pushed forward as a matter of urgency (cf. Pro-Rail Alliance 2024). European countries risk failing to meet their climate policy targets for the freight transport sector. Major core routes in Europe are congested due to massive truck traffic. The slow expansion of trunk roads contrasts with a large increase in transport performance by heavy goods vehicles (cf. Weidemann 2022). An increasing emphasis is being placed on the formulation of standards and guidelines to support sustainable transportation (cf. Curtis 2024). The trend towards the use of alternative drive systems in logistics is gaining momentum (cf. Kummer 2023).

European road traffic is threatened by a shortage of drivers, permanent traffic jams and the associated cost increases. Since 2010, road transport has already exceeded the capacity of the highways. The Brenner Highway, serving as the primary transit route between northern Italy, Tyrol, and Bavaria, ranks among the busiest traffic corridors in Europe. To mitigate carbon emissions, logistics networks along this axis must be optimized (cf. Kummer 2023). Over 13,000 trucks cross the Alps via the Brenner Pass every day. The congestion leads to massive traffic jams on both sides of the Brenner Pass. The federal state of Tyrol has introduced a truck traffic metering system to limit congestion on the route to 300 trucks per hour. Modeling serves as a vital tool for quantifying the impacts of sustainable transport strategies (cf. Wright 2023). All countries agree that the climate targets can only be achieved through a massive shift from road transport to environmentally friendly rail. Freight transportation by rail consumes 60% less energy and produces as much as 80% fewer emissions. Despite subsidies, rail's share of freight transport has been falling continuously for years (cf. Weidemann 2022).

Sustainable logistics requires the continuous optimization of transport processes and presupposes sustainable logistics and the comprehensive integration of ecological aspects into supply chain strategies (cf. Meißner 2024). This includes both ecological and economic efficiency (cf. Hofmann 2023). Sustainable logistics solutions require extensive research and development efforts (Hall 2023). Sustainable mobility requires the integration of environmentally friendly technologies into existing transportation systems (cf. Chiara 2023). Sustainable mobility in freight transport relies on emission-free vehicles and alternative drive systems (cf. McKinnon 2023). In Czech Republic roadside unit will be installed to make highways saver and smarter (cf. Yunex Traffic 2025). In France, road transport will remain the most important form of freight transport across the Alps in the long term.

3.4 Differentiation from intermodal maritime transport

As the backbone of international freight transport, sea freight manages the majority of global trade flows. Despite its economic importance, sea freight poses challenges to achieving environmental and social sustainability. Maritime transport is a central component of the global supply chain, with 77% of EU foreign trade and 35% of intra-European trade taking place by sea. Despite pandemic-related declines in 2020, growth is expected in the coming decades due to increasing demand for raw materials and container shipping (cf. EEA 2024).

A report by the European Environment Agency (EEA) and the European Maritime Safety Agency (EMSA) sheds light on the environmental impact of maritime transport. With a share of 13.5% of EU transport-related emissions, maritime transport is the third largest emitter after road transport (71%) and aviation (14.4%). Ship-related CO_2 emissions in EU ports amounted to around 140 million tons in 2018 (cf. EMSA 2024). In terms of air pollution, sulphur dioxide emissions amounted to 1.63 million tons in 2019, which accounts for 16% of global SO_2 emissions from maritime transport (cf. EEA 2024).

Sustainability is of paramount importance in maritime freight, particularly due to the significant impact of underwater noise pollution. Additionally, given the limited space available on a ship, the social component also becomes a critical factor.

Underwater noise is of particular importance in intermodal maritime transport. From 2014 to 2019, underwater noise caused by ships in EU waters doubled. The main sources are container ships, passenger ships and tankers. Around 50 % of the invasive species recorded in EU waters since 1949 are attributable to maritime traffic (cf. Racca et al. 2024).

Figure 7: Ecological consequences Maritime freight
Source: Author's own representation.

Since 2010, only three of the 18 major oil spills worldwide have occurred in EU waters, reflecting enhanced monitoring and heightened environmental awareness. Decarbonization of maritime transport requires innovative technologies and alternative fuels such as biofuels, hydrogen and ammonia. Onshore power supply can also significantly reduce emissions in ports (cf. Inal 2024).

EU Commissioners emphasize that shipping must become more sustainable and resilient to support the European Green Deal. The transformation of the sector is essential for the protection of fragile ecosystems, economic competitiveness and the creation of quality jobs (cf. von Malmborg 2024).

3.5 Interim conclusion

Freight transport constitutes a significant contributor to global GHG emissions. In 2023, freight transport accounted for around 19% of global GHG emissions, with road transport continuing to account for the dominant share in Europe at 77% (cf. Umweltbundesamt 2024). Despite political initiatives such as the European Green Deal, which aims to reduce emissions by 55% by 2030 and achieve climate neutrality by 2050, the shift of freight transport from road to rail remains sluggish. The low electrification of rail networks and the high CO_2 emissions associated with the construction and maintenance of transport routes pose additional challenges (cf. VDE 2022).

Sustainable transportation strategies in freight transport must encompass both technological innovations and political measures. They aim to minimize environmental impact while increasing efficiency. The integration of renewable energy and low-emission technologies is a central approach to increasing competitiveness and improving the quality of life in urban areas. Technologies such as the cargo tube, based on the Hyperloop concept, offer innovative solutions for reducing CO_2 emissions and energy losses in freight transportation.

Electrification and the use of hydrogen in freight transport are becoming increasingly important to replace fossil fuels and achieve climate targets. To meet its target of a 70% reduction in domestic transport emissions by 2030, Sweden is investing in electric truck deployment and the creation of green logistics corridors. The electrification of rail transport and the construction of hydrogen filling stations for trains are important steps towards reducing dependence on diesel and improving the CO_2 balance (vgl. Hydrogen Europe 2025).

A key factor for the efficiency of rail transport is the expansion and maintenance of the rail infrastructure. However, this is associated with high CO_2 emissions, especially when building new tracks and stations. The electrification of border crossings and the expansion of multimodal transport solutions are also crucial to promoting the shift of traffic from road to rail. Despite existing challenges such as differing standards and requirements in various countries, cross-border rail freight transport in Europe requires greater coordination and harmonization.

The future of freight transportation is closely linked to the use of green energy and sustainable technologies. Green logistics, defined by the systematic incorporation of sustainable measures throughout supply chains, is becoming the benchmark for the logistics industry. The use of hydrogen and battery EMUs as well as the electrification of the transport network are key technologies for reducing CO_2 emissions and improving the environmental footprint. Although hydrogen technologies present clear benefits, their efficiency is still in question due to significant energy losses in the electricity-to-hydrogen conversion process.

The sustainable transformation of freight transport is a complex challenge that requires a combination of political measures, technological innovations and infrastructural adjustments. The switch to low-emission and environmentally friendly transport solutions will be crucial to achieving global climate targets while increasing efficiency and competitiveness in European freight transport. The integration of green energy and new technologies such as cargo tubes and hydrogen trains offers considerable potential to reduce the ecological footprint and promote sustainability in logistics.

Maritime freight is indispensable to global trade flows, dominating international cargo volumes and constituting 77% of the EU's foreign trade and 35% of its intra-European trade. While its economic importance is indisputable, maritime transport poses significant environmental challenges. It is the third-largest contributor to transport-related emissions in the EU, responsible for 13.5% of the total, with CO_2 emissions in EU ports reaching 140 million tons in 2018. Additionally, ships emitted 1.63 million tons of sulfur dioxide in 2019, representing 16% of global SO_2 emissions from the sector. Underwater noise from shipping doubled in EU waters between 2014 and 2019, disrupting marine ecosystems, and maritime traffic has been linked to the introduction of 50% of invasive species in EU waters since 1949. Notwithstanding these issues, significant progress in risk management is demonstrated by the occurrence of merely three major oil spills in EU waters since 2010.

Decarbonization is critical for aligning maritime transport with sustainability goals and requires innovative approaches, including alternative fuels like biofuels, hydrogen, and ammonia, as well as onshore power supplies to reduce emissions in ports. EU policy-makers underline the necessity of transforming the sector to meet the European Green Deal's objectives, emphasizing that sustainability efforts will protect ecosystems, enhance economic competitiveness, and create quality jobs.

4 Research design and empirical validation

This chapter centers on the practical implementation of sustainable mobility. It outlines the research design in detail and formulates the resulting research questions.

This study places particular emphasis on maintaining a high level of methodological quality, adhering to established scientific standards in the social and economic sciences. As part of the operationalization process, guiding questions and analytical categories were systematically developed based on the central research questions. Prior to data collection, precise main and sub-questions were formulated to structure the interviews and guide the focused analysis.

The empirical survey of the dissertation is carried out using the qualitative research approach. The qualitative approach aims to answer the research questions and, on this basis, to make optimization recommendations for European freight transport. The research questions could not have been fully answered by quantitative methods, which is why the quantitative research approach, and the mixed-methods approach were rejected after careful consideration by the author.

The author initially examines foundational methodological approaches and consciously chooses a qualitative design over a quantitative one, given its superior suitability for exploring a limited number of cases in depth.

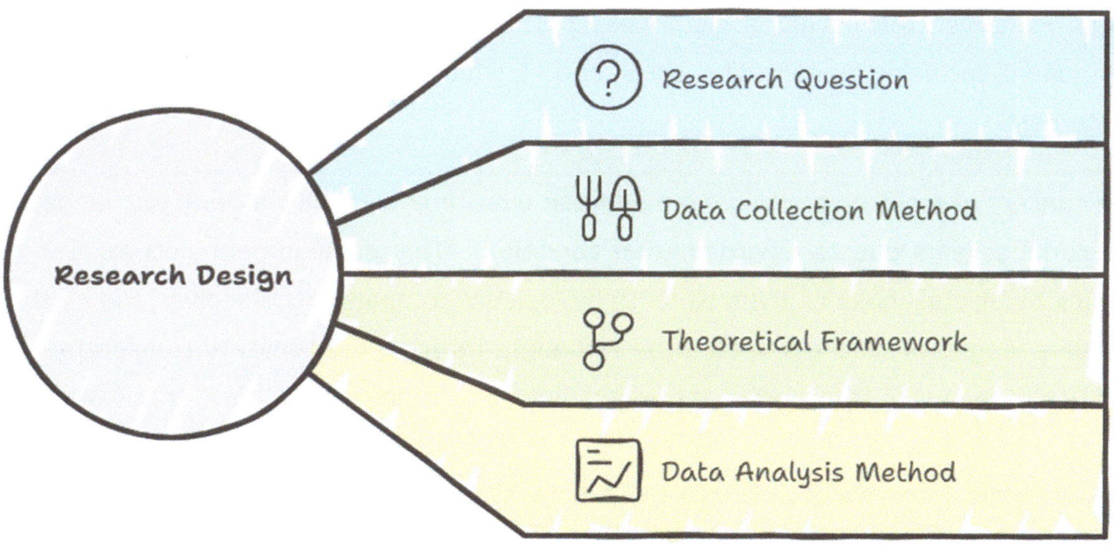

Figure 8: Qualitative evaluation method
Source: Author's own representation.

The chosen methods are based on existing scholarly knowledge and follow a transparent and rule-guided approach to ensure intersubjective traceability and a sound scientific foundation.

4.1 Literature analysis as the basis for the design

To ensure a basis for empirical validation and research, the basics of mobility, environmental goals and freight transport in Europe are extensively researched in advance. In addition to traditional research methods involving libraries and printed literature, digital resources such as Google Scholar, EBSCO, and ResearchGate are also utilized to narrow the broad scope of freight transportation and to establish a solid technical foundation for the study. Beyond academic journals, the literature review encompassed scientific articles and scholarly essays to ensure a comprehensive analysis. The PRISMA analysis with the flow diagram is used to identify the correct literature sources and for further evaluation, see figure in Appendix 25.

Key constructs such as green energy and the European Green Deal are defined and a distinction is made between clean, green and sustainable energy. The classic three-pillar model of sustainability is supplemented by a further pillar, the technological pillars, as technology and the associated opportunities will be essential in 2025. Technology and information systems, particularly artificial intelligence (AI), are expected to play a decisive role in this context, with their influence likely to grow significantly in the coming years. Moreover, this insight is the novelty of this work, as it is precisely this technical component, including digitalization issues and AI, which has not been examined in depth in the past.

The concept of mobility has also been adapted over time and will continue to change in the coming years due to environmental conditions. This development necessitates a rethink by all stakeholders involved, both by logistics companies themselves and by the shippers. Logistics concepts need to be rethought to avoid unnecessary transportation and use of means of transport. A paradigm shift is currently taking place that is having a profound impact on all those involved. In addition to traditional sources, trends were also analyzed to derive a view into the future. The research questions for the expert interviews were derived based on the literature and current trends. Before empirical validation can begin, the literature analysis serves as a sound basis and provides basic information for the further procedure. The next step is to carry out an environmental analysis in preparation for the expert interviews. Based on the results and findings of these expert interviews, the second part of the SWOT analysis is carried out, which clearly highlights the future-oriented strengths and weaknesses.

4.2 Presentation of the SWOT analysis

The term SWOT is an acronym for *Strengths, Weaknesses, Opportunities,* and *Threats*. These four dimensions are typically presented in a matrix format. The purpose of the SWOT analysis is to assess internal strengths and weaknesses in relation to external opportunities and threats in the business environment. It serves as a strategic tool for comparing a company's internal and external influencing factors in order to support informed decision-making (cf. Asana 2025). An objective comparison opens opportunities for a goal-oriented and conscious design of core competencies and a strategic orientation.

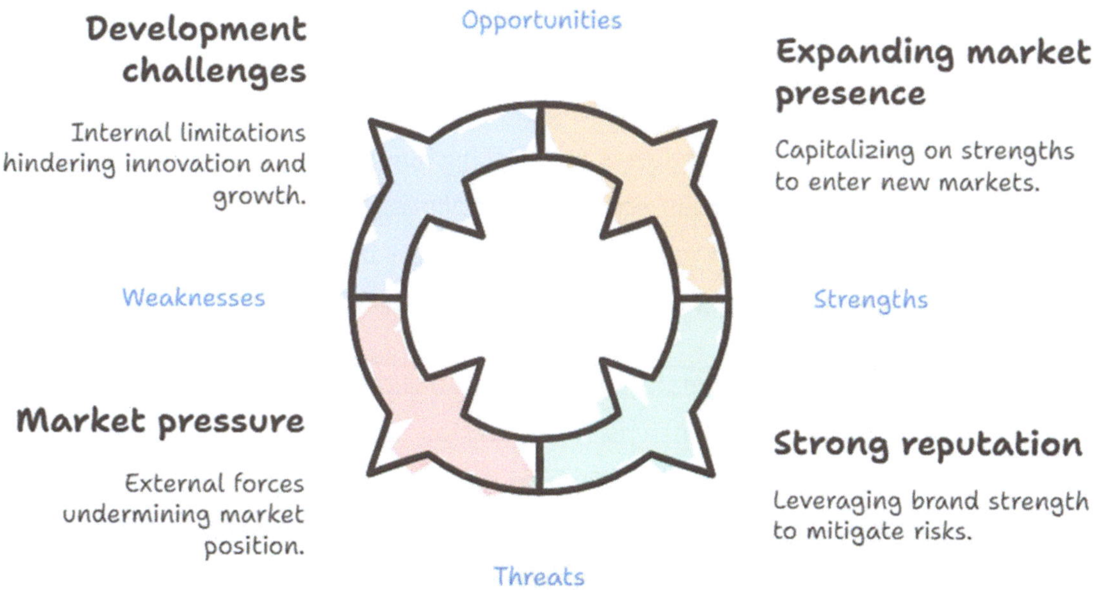

Figure 9: Template SWOT analysis
Source: Author's own representation.

This method is suitable for strategic planning, positioning and strategic management. In companies, the SWOT matrix serves as a starting point for developing strategies that build on the identified strengths, address weaknesses, exploit opportunities and minimize risks (cf. Asana 2025). The strengths and weaknesses of the market are drawn up based on existing literature. Chapters 5.3 and 5.4 visualize the opportunities and risks that are geared towards the future. This evaluation is based on the findings of the expert interviews.

A SWOT analysis is employed as an analytical framework to structure and synthesize the findings in terms of strengths, weaknesses, opportunities, and threats. The integration of this strategic analysis tool aims to present the results in a practice-oriented manner and to interpret them with a view toward future developments. This approach enables a structured and focused examination of the research questions related to the opportunities and risks associated with the "mobility competition". The combination of arena analysis, expert interviews, and SWOT analysis is intended to first theoretically delineate relevant stakeholders and thematic areas, then empirically gather expert perspectives, and finally interpret these insights within a practice-relevant analytical framework.

4.3 Derivation of the causal model

A causal model serves as a structured instrument for analyzing cause-and-effect relationships between different variables (cf. Plan Concepts 2025). A central aspect of this model is to distinguish between causality and mere correlation. Such relationships are visualized using graphical representations, particularly in the form of path diagrams. The causal model is not only a theoretical concept, but also a methodological tool for hypothesizing and testing causal hypotheses (c.f. Hochschule Fulda 2014). The application of causal models spans numerous scientific disciplines, offering valuable insights into complex relationships between variables. They support informed decision-making and enable the empirical testing of theoretical assumptions.

Causal models are powerful tools in data analysis that facilitate the understanding and prediction of cause-and-effect relationships. They are founded on the assumption that changes in one or more causes directly influence corresponding outcomes. By using causal models, it becomes possible to analyze complex systems and make well-informed decisions by identifying and quantifying the underlying mechanisms driving observed phenomena. The following causal model, which is relevant and valid for this work, results from the methodological approach and the theoretical basis. The causal model used in this work is based on theoretical foundations and tailored to its specific context (cf. Plan oncepts 2024).

The presented causal model illustrates the interrelationships between various dimensions of sustainability and their influencing factors. Sustainability is divided into economic, environmental, technological, and social dimensions, each contributing to and interacting with central factors. Green energy focuses on promoting renewable energy sources to reduce CO_2 emissions, while digitization leverages technological innovations to optimize processes and increase efficiency.

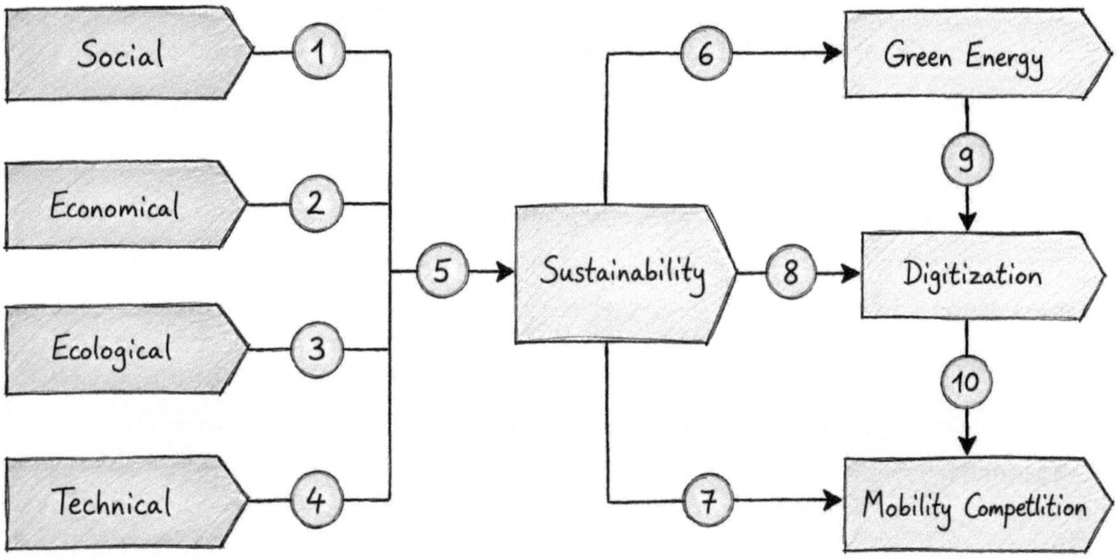

Figure 10: Causal model
Source: Author's own representation.

The diagram presents a causal impact model of sustainability. It illustrates how various influencing factors affect the central construct of *sustainability*, which in turn leads to downstream effects in domains such as *green energy*, *digitalization*, and *mobility competition*. The numbered circles represent sequential steps within the causal chain.

Input Variables (Steps 1–4): Positioned on the left side of the diagram, the following four dimensions represent the key exogenous drivers of sustainability:

1. Social Dimension: Encompasses societal factors such as education, equality, and social responsibility, which significantly influence sustainability outcomes.
2. Economic Dimension: Includes elements such as economic stability, investment readiness, and market-driven mechanisms that foster sustainable development.
3. Ecological Dimension: Refers to environmental awareness, resource conservation, and climate-related objectives, which serve as fundamental ecological drivers of sustainability.
4. Technological Dimension: Captures the role of innovation, digital technologies, and technical efficiency as enablers of sustainable transformation.

Central mechanism:

5. These four input dimensions collectively shape or promote sustainability as the central endogenous variable in the model. This convergence occurs in the middle section of the diagram.

Primary Outcomes of Sustainability (Steps 6–8):

6. Impact on Green Energy: Sustainability promotes investments and strategies in the field of renewable energy and supports the transition toward low-carbon energy systems.
7. Impact on Mobility Competition: Sustainability alters market dynamics in the mobility sector, for instance by encouraging the adoption of emission-free transportation technologies.
8. Impact on Digitalization: Sustainable strategies drive digital transformation, such as the development of smart grids, circular digital economies, and Internet of Things (IoT) solutions.

Secondary Causal Chain (Steps 9–10):

9. Green Energy → Digitalization: The expansion of green energy necessitates advanced digital control systems and smart infrastructure solutions.
10. Digitalization → Mobility Competition: Digital technologies enable new business models in the mobility sector, e.g., car sharing, autonomous vehicles, thereby reshaping competitive conditions.

4.4 Empirical Investigation

A thorough analysis of the environment is crucial for the structured planning of expert interviews, as it enables the identification of the most relevant stakeholders to be interviewed. An arena analysis is carried out for this purpose. Upon completion of the arena analysis, the guided expert interviews are systematically planned and carried out. Drawing on the findings from both research methods, a SWOT analysis is performed to systematically identify internal strengths and weaknesses alongside external opportunities and threats. This enables the formulation of qualified recommendations for action. The expert interviews will be conducted between July and November 2024.

The interaction between the individual areas appears to be important to be able to assess the organization as a whole. This analysis can be broken down to the organization or individual organizational units, allowing conclusions to be drawn about the company.

At the outset, an arena analysis is conducted as a preparatory environmental and stakeholder analysis to systematically identify relevant actor groups and thematic focal areas for the subsequent expert interviews. This preliminary analysis ensures that the selection of interview participants is based on a well-founded and objectively justified rationale, aligned with the key actors and core topics of the research field under investigation.

4.4.1 Derivation of the research questions

In the previous elaboration of this work, it became clear that freight transport in Europe will increase sharply in the coming years, so a massive increase in CO_2 emissions and other nitrogen oxides of up to 45% is also expected. Simultaneously, efforts are focused on meeting climate targets that require a 55% reduction in CO_2 emissions (cf. Fraunhofer ISE 2024a). In future, more traffic will be transported by rail, as the theoretical research in the previous chapters has already shown.

To ensure compliance with the climate targets, the central research question and three sub-questions are posed. All four questions are to be answered in the subsequent interviews and analyses in accordance with the resulting findings:

Key question

What are the potentials and risks of mobility competition?

Due to the constantly changing requirements of the market, logistics companies are forced to adapt to these requirements or even offer pioneering solutions for customers. The transformation of mobility presents both opportunities and risks for future actions. The potential and risks are identified and discussed in the further course of the thesis. To provide a clear answer to the central research question and to derive recommendations for optimizing mobility competition, three supporting sub-questions have been formulated.

Sub-questions

(1) To what extent must European freight transport adapt to achieve the sustainability goals by 2040?

The first sub-question aims to clarify what European freight transport needs to do specifically to achieve the sustainability targets that have been set. Although there are still approximately 15 years until 2040, the scope and scale of measures should be defined as soon as possible, since the challenge of rising transport volumes must be addressed alongside the achievement of climate targets.

(2) *What specific measures can help to make European freight transport more sustainable in line with the climate targets that have been set?*

The second sub-question is about the specific measures. Nevertheless, measures must not only focus on optimizing outcomes and meeting targets but also ensure sustainability to achieve lasting success. Furthermore, the targets should be aligned and financially viable.

(3) *How can green energy be used sensibly in European freight transport to achieve sustainability goals?*

The third sub-question is about the sensible use of green energy. Green energy plays an important role in achieving climate targets when considering renewable energy. The significance of this role will continue to rise in the coming years. To remain successful in the marketplace, companies must actively address and integrate green energy solutions.

4.4.2 Quality criteria for scientific research

To ensure that the scientific approach can be recognized as such, three quality criteria must be met. The reliability of the collected knowledge is an important asset. Figure 11 shows the three principles (cf. Gläser / Laudel 2010):

- The first quality criterion describes the openness within the subject area to be researched. This implies that the scientific process must unconditionally incorporate and accept new information that emerges during the course of the investigation (cf. Mayring 2015). In particular, the acceptance and classification of contradictory information is essential. In addition, during research it is expected that new information is not prematurely pushed into pre-formulated categories, which could result in new information being overlooked (cf. Mayring 2022).

- The second criterion describes the consideration of previous theories and scientific findings. To be able to contribute to the overall state of knowledge with the expected increase in knowledge through the study, the research approach should therefore build on previous knowledge. The comprehensibility of the chosen approach also plays a major role (cf. Gläser & Laudel 2010).

- This can be ensured by observing and naming explicit rules during the investigation. In the first step, a high level of transparency is essential. With the help of a detailed description of all research steps, trust in the study and the resulting findings can be established.

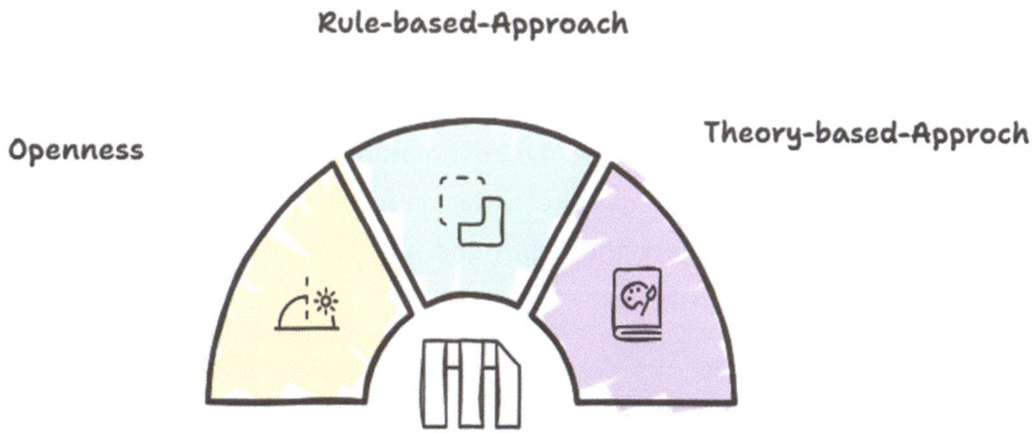

Figure 11: Principles of research
Source: Based on: Gläser/Landel 2010.

At the same time, this makes it possible to discuss the classification of the results in a scientifically sound manner (cf. Gläser & Laudel 2010). The following chapter describes the selection of methods for the study. The identified quality criteria are systematically integrated and rigorously applied throughout the research process.

4.4.3 Selection of the methodology

In order to take a closer look at European freight transport with regard to sustainability and the use of green energy, the scientific approach must first be clarified. It is important to define the empirical approach. The distinction between a quantitative, qualitative or mixed-methods approach is fundamental. The quantitative approach has its origins in the natural sciences. The research entails the utilization of a comprehensive statistical database, compiled through the standardized and systematic collection of data.

The aim of the quantitative approach is the falsification or confirmation of theory-based hypotheses and the verification of previously established assumptions. The qualitative approach works with considerably less data and is usually based on interviews or observations. The focus here is on a detailed investigation of a small number of cases and the clarification of research questions. One advantage of this approach is the flexibility during the investigation. Preliminary optimization measures are formulated based on the insights gained from the interviews conducted and their subsequent analyses.

The use of expert interviews is considered an appropriate method, as the underlying research questions aim to access specialized experiential knowledge that is best obtained through direct engagement with qualified interviewees. Semi-structured interviews guided by an interview protocol are employed to ensure both comparability across interviews and openness to emerging, unanticipated insights. This methodological approach offers the advantage of collecting verbal data in a flexible and context-sensitive manner, an essential feature given the dynamic nature of the subject area, namely sustainable mobility and green energy. Alternative data collection methods are implicitly deemed less suitable, as they lack the depth and adaptability required to adequately explore future potentials and risks.

4.5 Presentation of Arena analysis

An arena analysis is suitable for identifying the right stakeholders and developing a solution for optimizing the sustainability of European freight transport. The arena analysis serves to investigate the contextual environment of companies or sectors by identifying key actors, defined as stakeholders who are directly or indirectly affected. Relevant issues are conceptualized as key topics and subsequently positioned within a model that classifies them according to their importance to the central theme. (cf. Pfaff 2020). Simultaneously, the identification of relevant stakeholders allows for the derivation of expert categories that are essential for the topic and the subsequent empirical investigation (cf. Osztovics 2021: 3).

Stakeholders (actors)	Employee survey topic areas (issues)
Company: Demand for more quality and sustainability and expansion of IT	Topics that ensure the company's own continued existence and growth in the markets
Supply chain manager: supply chains and resource management	KPI management topics to optimize the value chains within the company
IT Manager: Leaders in the IT environment, CIO etc.	Topics relating to IT, security and future-oriented IT and AI solutions
IT provider: Provision and operation of relevant IT infrastructure	Topics relating to the replacement of existing systems and the expansion of new market functions
Market researchers: knowledge carriers about new trends	Topics for analyses and surveys to derive trends
Logistics: know-how carrier for all transport areas and processes	Connection to the systems for process optimization of transport times and cost optimization
Infrastructure: Experts for planning infrastructure issues	Topics such as the creation of new or modernization of existing infrastructure, in terms of roads, charging facilities, etc.

State: Represented by ministries, e.g. Ministry of the Environment	Political issues to ensure an orderly framework
Associations: Associations such as the VDV or Pro-Rail Alliance	Socio-political issues and regulations that are resolved for companies at a higher level
Software manufacturers: Leading software manufacturers of modern and innovative software	Topics such as the implementation and programming of modern IT solutions to carry out the activities
Management boards: Decision-makers such as MD, CEO, CFO etc.	Issues that serve the profitability and continued existence of the companies

Table 4: Stakeholders and issues in European freight transport
Source: Author's own illustration based on Pfaff 2020.

The arena analysis comprises both stakeholder and issue analysis. Stakeholder analysis focuses on identifying individuals and groups that position themselves, or are positioned by others, as key actors within the arena. The issue analysis is the second part of the arena analysis and examines the demands and relevant employee survey topic areas of the market participants. The stakeholders involved in optimizing the sustainability of European freight transport vary. Table 5 lists the stakeholders most closely associated with the topic of optimizing the sustainability of freight transport. In the context of the application, stakeholders are added in the form of logistics companies and associations, which can be categorized. Table 5 can be summarized in four categories.

Once the relevant actors have been identified, the next step involves determining the issues these stakeholders are confronted with. A wide variety of issues arise during implementation, design and use, which can be condensed into the factors of quality and safety. Issue analysis is the identification, analysis and evaluation of relevant social topics (issues) with high relevance.

The company	Rollers	market (supply chain)	Regulations
Transportation company	CIO, MD, CEO	Large companies	Politics + State
Associations	Authors	Software	European Union
Specialists	Market researcher	Logistics	Technology
IT solutions	Experts	Infrastructure	Infrastructure

Table 5: Categorization of the identified stakeholders
Source: Author's own representation.

While these issues have garnered considerable attention within expert communities, they remain largely absent from public discourse. Consequently, this study identifies and examines these issues through systematic surveys of technical specialists. To facilitate visualization and subsequent analysis, the stakeholder clusters are mapped within a positioning framework, with axes determined by the factors identified through the issue analysis (cf. Pfaff 2020). The extended environmental analysis shows all relevant stakeholders that need to be considered and included. This diagram is used to assign the identified stakeholders to the individual environments and to visualize them. In the *company* area, specific companies are named that are affected, i.e. the transport and logistics companies themselves as well as the customers who send and receive the freight.

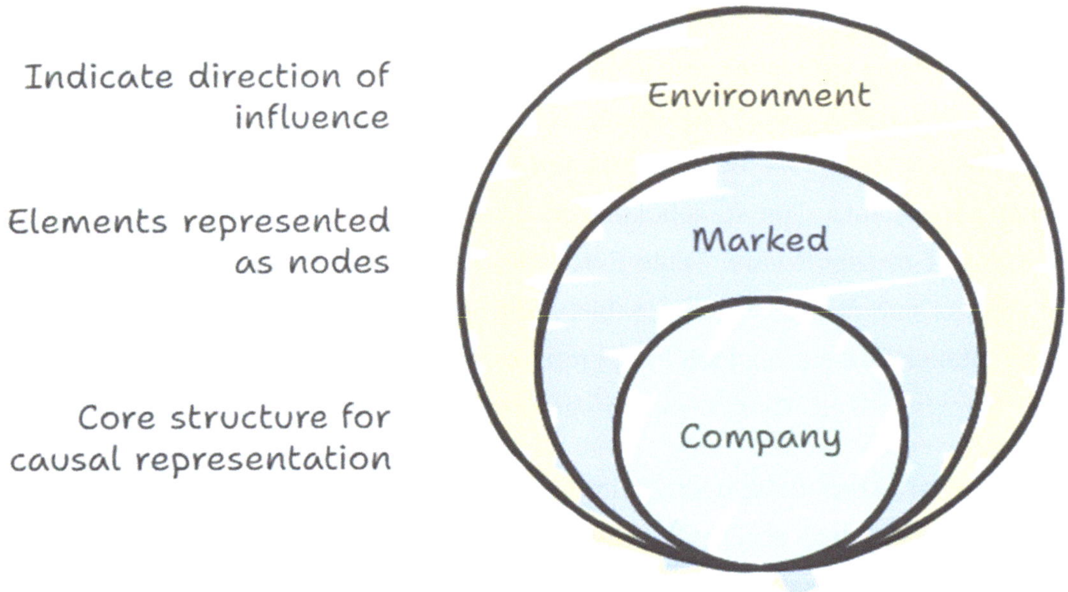

Figure 12: Template of the extended environment analysis
Source: Author's own illustration based on Pfaff (2020).

The *market* as a superordinate area shows the competition and other logistics companies, including companies such as OEMs and the like. The surrounding area *environment* describes the external environment such as the state, associations and institutes. It is also important to consider the impact of political decisions and international trade agreements on freight transport and emissions. The involvement of environmental organizations and non-governmental organizations can help to promote more sustainable practices and ensure compliance with climate targets. This extended environmental analysis provides a systematic overview of key stakeholders in freight transport, examining their individual and collective impacts on climate target attainment.

4.6 Presentation of expert interviews

4.6.1 Interview planning

This chapter employs a qualitative approach, building upon the substantial foundational data established through the preceding environmental and SWOT analyses. However, there is still a lack of concrete information, impressions and plans to support the sustainability of European transport companies from an internal perspective. Given the complexity of this subject area, a comprehensive understanding of underlying context is essential. The qualitative approach facilitates the collection of nuanced, in-depth information that would be inaccessible through quantitative methods alone. (cf. Dresing and Pehl 2018: 68).

Furthermore, it is not possible to obtain the required information through research and analysis, as this is a broad field of research with a focus on European freight transport (cf. Brüsemeister 2008). The inherent flexibility of qualitative research methodology permits iterative refinement of the empirical study design based on insights gained during initial data collection phases (cf. Brüsemeister 2008). This procedure ensures adherence to established quality criteria for scientific research.

Figure 13: Planning the interviews
Source: Author's own representation.

A great deal of information on sustainability in the context of sustainability and green energy was collected in advance and is now being incorporated into the study. To capitalize on this methodological flexibility, semi-structured expert interviews were selected as the primary data collection approach. This is a semi-structured approach in which verbal information is recorded/collected (cf. Misoch 2019). To structure and evaluate this information, the content analysis was carried out using qualitative content analysis according to Philipp Mayring. Before starting, research questions were formed based on the research. These serve as a focus during the interviews and the subsequent analysis.

Once the methodological procedure has been described and concretized, the interview process must be outlined. The planned steps are shown in the diagram. The initial step involves developing comprehensive interview guidelines that address several essential framework conditions. Suitable interview partners must then be found. Once the experts have been identified, the interviews are planned and conducted. These interviews are recorded (audio stream). They are then recorded in writing and transcribed. The expert interviews are controlled using an interview guide, which was drafted in advance of the study with the help of collected theory. In addition to listing all the technical questions, the interview guide is also used to structure the interview. The interview protocol must therefore commence with an information phase, during which participants are briefed on the research objectives and the intended use of collected data.

Clear, content-driven selection criteria were defined to recruit a relevant and diverse group of experts. Expert selection was governed by four primary criteria: established sustainability engagement, domain-specific specialization, operational experience within freight transport, and demonstrated green energy expertise. This ensures that the selected individuals possess both subject-matter expertise and the ability to contribute varied perspectives on the research topic. The preceding arena analysis served as the conceptual basis for this selection by systematically identifying key stakeholder groups. The selection strategy aimed to encompass diverse sectoral representation to ensure comprehensive perspective capture across the field. This deliberately heterogeneous selection follows the principle of theoretical sampling, which aims to explore the research field in its full thematic breadth rather than relying on a random sample.

Consent is also obtained at this stage so that the subsequent interview can begin (cf. Misoch 2019). For the actual interview, the guidelines are analyzed by a working group and checked for application. To ensure the quality of the data collection instrument, two pretests of the interview guide were conducted. Based on the results of the first pretest, targeted adjustments, such as rewording, additions, or reordering of individual questions, were made and directly integrated into the revised guide. A second pretest was then carried out to reassess the effectiveness of these modifications. This iterative approach reflects methodological reflexivity and contributes significantly to the optimization of the data collection instrument. Pilot testing validated both the structural coherence of the interview protocol and its effectiveness in eliciting theory-driven insights, significantly strengthening data collection reliability.

The interviews were conducted via video conferencing (Zoom or Microsoft Teams) to enable personal interaction despite physical distance. This form of communication ideally promotes participants' willingness to cooperate, and fosters trust in both the interviewer and the research process. All interviews were conducted under standardized conditions to maintain methodological consistency throughout the data collection process.

Consent to the use of the interview recording is requested at the beginning of the interviews. The interview is then recorded digitally. To ensure that the recording is fail-safe, two smartphones are used in parallel. Once the interview has been completed, it is transcribed. The most important rules for the preparation and transcription are based on Dresing and Pehl (cf. Dresing & Pehl 2018). Once the interviews have been conducted, they are analyzed using Mayring's qualitative content analysis. A nominal category system is used within the analysis. In advance, superordinate categories are deductively established based on the theoretical analysis (cf. Mayring 2022).

All interviews underwent verbatim transcription with colloquial expressions standardized, punctuation normalized, speakers identified through abbreviations, and sequential line numbering applied. This transcription and coding process forms the basis for a reliable analysis. Adhering to such standards increases the transparency of data interpretation, as it becomes traceable how raw data were transformed into interpreted categories. The interview questions were deliberately formulated in an open-ended manner to allow experts the space for unrestricted responses and to avoid missing potential insights through overly structured or closed questions. This approach actively maintained openness, a fundamental quality criterion of qualitative interview methodology.

4.6.2 Creation of the interview guide

This chapter provides a comprehensive examination of the interview guide structure and content. The interview guide can be found in Appendix 3. The guide should fulfill the following characteristics:

1. Focus on the subject area
2. Highlighting the most important employee survey topic areas
3. Creating comparability by sorting the topics
4. Ensuring an orderly flow of conversations

Questions must be structured as open-ended inquiries to allow participants maximum response flexibility and ensure authentic perspective capture. This methodology ensures comprehensive insight capture by avoiding the analytical limitations imposed by closed question formats (cf. Misoch 2019). In addition to listing all the technical questions, the interview guide is also used to structure the interview. First, the guide should contain the so-called information phase. In this phase, the interviewee is informed about the objective of the interview and how the interview content will be used. Consent is also obtained in this phase so that the subsequent interview can begin. The introductory phase of the interviews is outlined in Table 6.

The technical introduction and presentation also take place at this point (cf. Misoch 2019).

No.	Questions	Interview phase	Category
A.1	- Gender: - Country: - Age: - Experience in years:	Entry phase	--

Table 6: Questions for the introductory phase
Source: Author's own representation.

The second phase also serves as an introductory phase to the conversation. It is advisable to ask a very open question with a wide range of answers. Such protocols enhance participant confidence and optimize engagement during the structured interview process (cf. Misoch 2019).

This is followed by the concrete formulation of the prepared questions. This can be done deductively and questions based on researched knowledge can be asked. It is also possible to derive questions inductively from answers during the interview (cf. Misoch 2019). A mixed approach was chosen for the following interviews. First, theoretically derived questions are asked. Spontaneous questions emerging during data collection are addressed following the completion of each thematic module, ensuring comprehensive coverage while maintaining protocol structure. Question adaptation is essential given the diverse expert backgrounds and specialized domain knowledge represented across participants.

The first interview section of the main phase deals with the potential and risks regarding a sustainable decarbonization of transport. The previous chapters have already looked at the transport sectors and analyzed them from a sustainability perspective. This part of the interview is intended to identify the potential and risks.

No.	Questions	Interview phase	Category
B.1	Where do you see the greatest potential for increasing sustainability?	Main phase	Potentials and risks
B.2	Where do you see the greatest risks in increasing sustainability?	Main phase	Potentials and risks

Table 7: Questions on potentials and risks
Source: Author's own representation.

The initial mapping of potential and risks provides the foundation for subsequent detailed analysis of comprehensive sustainability considerations as shown in table 8 below. In the previous chapters, the scope of energy consumption was identified as an important component of a sustainable decarbonization of transport. As sustainability is also very interesting from a social perspective, the following two questions on social sustainability aspects are addressed.

No.	Questions	Interview phase	Category
C.1	How can European freight transport ensure environmental sustainability in the future?	Main phase	Ecological Sustainability
C.2	What needs to be improved in the construction of infrastructure and the development of drives?	Main phase	Ecological Sustainability

Table 8: Questions on environmental sustainability
Source: Author's own representation.

The objective is to identify necessary measures for European freight transport to maintain employee motivation and enhance workforce qualification for emerging industry requirements. The second inquiry examines the incentive structures that policymakers and corporations must establish to foster or enhance future social acceptance.

No.	Questions	Interview phase	Category
D.1	Which aspects of social sustainability will play a greater role in European freight transport than they do today?	Main phase	Social Sustainability
D.2	What measures and incentives could there be to increase social acceptance in the future?	Main phase	Social Sustainability

Table 9: Questions on social sustainability
Source: Author's own representation.

A growing market share in the transport sector is key to economic sustainability. Therefore, the interview will focus on the economic aspect. As transport networks are already at full capacity, mobility and new mobility concepts are becoming increasingly important.

No.	Questions	Interview phase	Category
E.1	How can freight transport ensure economic sustainability?	Main phase	Economic Sustainability
E.2	The infrastructure has already reached the limits of its capacity. What economic measures could relieve the strain on the infrastructure in the future?	Main phase	Economic Sustainability

Table 10: Questions on economic sustainability
Source: Author's own representation.

Technical sustainability constitutes an additional thematic block within the interview framework. With the demands of the market, technical aspects are also being discussed, which are becoming increasingly important. These two broad-scope questions encompass both specific freight transport technologies and emerging areas such as artificial intelligence and digitalization.

No.	Questions	Interview phase	Category
F.1	How could technology change in the coming years?	Main phase	Technical sustainability
F.2	How could IT or artificial intelligence solutions increase sustainability?	Main phase	Technical sustainability

Table 11: Questions on technical sustainability
Source: Author's own representation.

Green Energy is another issue that will become increasingly important in the coming years. E-mobility assumes increasing significance not only in automotive transportation but also within commercial freight transport sectors. This thematic block explores alternative propulsion technologies alongside policy instruments intended to accelerate sustainable energy utilization. This block also questions political influences and what can be done to produce as little greenhouse gas as possible in the future.

No.	Questions	Interview phase	Category
G.1	What incentives could politicians create to increase the use of green energy?	Main phase	Green Energy
G.2	How do infrastructure, technology and drives need to change to produce less greenhouse gas for future generations?	Main phase	Green Energy

Table 12: Green Energy
Source: Author's own representation.

To find a suitable conclusion to the interview, the expert is given the opportunity to raise any outstanding points on the entire topic. This procedure also serves to officially end the interview situation for both sides (cf. Misoch 2019). The following question is asked at the end of the interview.

No.	Questions	Interview phase	Category
H.1	Are there any other points you would like to talk about? Thank you very much!	Final phase	--

Table 13: Derivation to questions of the final question + other topics
Source: Author's own representation.

This means that 13 questions are asked, which are based on the guidelines. If necessary, the questions can be supplemented by further intermediate questions or by questions from the interview participants. Here are several improved versions at dissertation level. Clear and methodologically focused: Demographic information, including professional role and age, is collected preliminarily, though analysis of these personal data falls outside the scope of this study.

4.6.3 Implementation

Participant identification and selection utilize standard market-available professional networking platforms and databases. Xing, LinkedIn and company homepages can be used to filter and search for specialist areas of individual companies and specialists. Particular attention is paid to the following characteristics during the search:

1. Experience in freight transport
2. Focus on sustainability issues
3. Expertise in driving and/or energy carrier topics
4. Expertise in green energy topics
5. Mobility

The aim is to interview experts from as many different but still relevant areas as possible by taking all characteristics into account. The following experts were recruited for the interviews. A total of 13 experts were found and interviewed. The 13-interview sample was systematically selected to capture the essential sustainability dimensions pertinent to the research objectives.

No.	Experts	The company
1	Experte 1	DHL
2	Experte 2	ObOvo
3	Experte 3	VDV
4	Experte 4	Deloitte
5	Experte 5	Pro-Rail Alliance
6	Experte 6	Schenker Rail
7	Experte 7	HySolutions
8	Experte 8	German Railways
9	Experte 9	CGI
10	Experte 10	Fraunhofer Institute
11	Experte 11	Swiss Federal Railways
12	Experte 12	Austrian Federal Railways
13	Experte 13	Brenntag

Table 14: List of experts for interviews
Source: Author's own representation.

Linguistic alignment between interviewer and participant is essential, requiring adaptive communication strategies to match participant language registers. In addition, the interviewer will keep the questions simple and only ask one question at a time to ensure comprehensibility. As established in the preceding analysis, the interview guide serves to provide thematic direction and structural framework rather than rigid procedural constraints. This results in a procedure that is as flexible as possible and avoids *fixed* questions on employee survey topic areas (cf. Misoch 2019). The interviews are conducted exclusively using digital means of communication, as the experts interviewed are based in many different locations throughout Europe. In all interviews, video telephony is used to facilitate *more personal contact*. Ideally, this increases cooperation and trust in the interviewer and his questions (cf. Scholl 2018).

4.7 Transcription and evaluation

Consent to the use of the interview recording is requested at the beginning of the interviews. The interview is then recorded digitally. To ensure that the recording is fail-safe, two smartphones are used in parallel. Once the interview has been completed, it is transcribed. This makes it possible to record the verbal questions and answers in writing so that the data can be used for later analysis. It is important to strike a balance between a high level of detail and appropriate implementation. The transcription approach prioritizes content-semantic accuracy over paralinguistic features, given that elements such as emphasis, dialect, and temporal pauses fall outside the study's analytical requirements. There are a few rules that must be observed when using this type of transcription.

The most important rules for the present elaboration are presented below (cf. Dresing & Pehl 2018):

1. The transcription is verbatim
2. Abbreviations and slang are formulated in writing
3. Punctuation is smoothed to ensure good readability
4. There is a separate paragraph for each speech
5. The interviewer is identified with a "KHD#", the expert with the respective initials "XY#"
6. Each line of the transcript is marked with a counting number

The transcripts created are attached to this work in Appendix 5. The list of complete transcripts and the audio files are included as an appendix to this paper.

4.7.1 Generalization, paraphrasing and condensation

After transcription, the expert statements are paraphrased and reduced for the first time. This reduction is based on the respective question blocks. This first reduction is shown in Appendix 21. This is followed by generalization and reduction 2 based on the individual questions. The research questions are addressed through this analytical generalization process. This second reduction is shown in Appendix 22.

The study is limited to freight transport within Europe. Public passenger transport and private journeys by car and other means of transport are not included in this study. Since conducting expert interviews is a research method that is in the interest of the researcher, social rules and compliance with data protection laws apply. Once the interviews have been conducted, a qualitative content analysis according to Mayring is carried out. A central component of the evaluation methodology is the continuous focus on the text and the respective context in which the statement was made. The material used for this content analysis was described in the previous chapters. The data corpus comprises exclusively individual interview transcriptions.

As the focus is on highlighting the most important information from the interviews and the subsequent SWOT analysis, the first procedure is therefore used in the following evaluation. In this dissertation, the analysis will begin with the earliest interview conducted and proceed chronologically to the most recent. An essential component of content analysis is the development and application of categories for analyzing the text (cf. Mayring 2022). A nominal category system is used in this analysis. In advance, superordinate categories were established deductively based on the theoretical analysis (cf. Mayring 2022). After the initial establishment of the categories, individual steps are taken to compress the text material.

4.7.2 Quality criteria according to Mayring

Mayring's quality criteria relate to qualitative research and qualitative content analysis. The expert interviews are evaluated according to Mayring's quality criteria, which are applied specifically to qualitative research:

1. Procedural documentation: Documentation includes details of the research process and methodology that enable other researchers to understand and, if necessary, reproduce the procedure.
2. Argumentative interpretation validation: Refers to the fact that interpretations are underpinned by arguments. Care is taken to ensure that the conclusions are presented convincingly.
3. Proximity to the subject: The research should be closely linked to the phenomenon under investigation. The aim is to take appropriate account of the perspectives of the people involved.
4. Communicative validation: means that the results are discussed with the experts in order to check the validity of the interpretation.
5. Triangulation: Refers to the use of different methods to validate the results, which leads to robustness of the research.
6. Intersubjective comprehensibility: The research results should be comprehensible to other researchers. This requires a clear presentation of the research steps and results.

Philipp Mayring, a prominent scholar in qualitative content analysis, developed quality criteria to ensure the scientific rigor and reliability of qualitative research. His approach emphasizes systematic, transparent, and reproducible analysis of textual data.

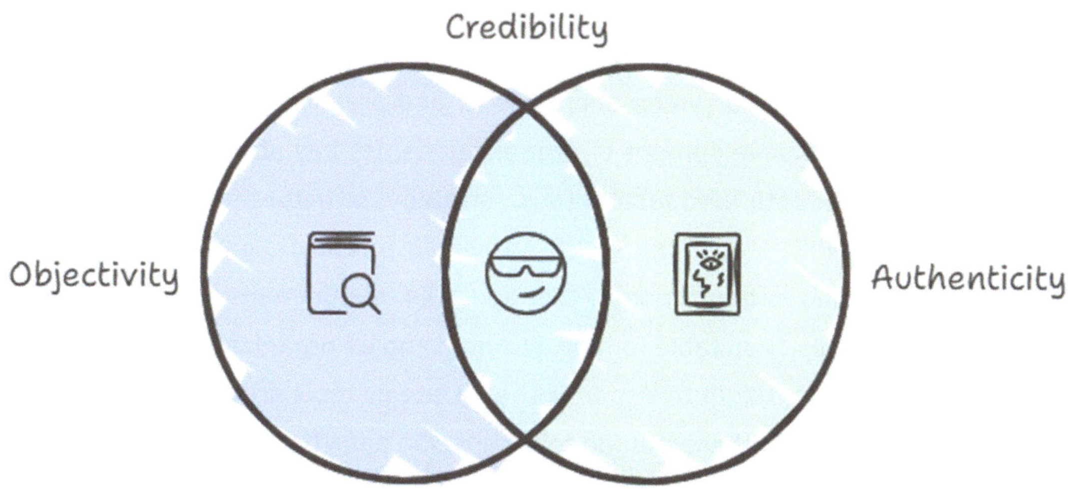

Figure 14: Quality criteria according to Mayring
Source: Author' own presentation.

The qualitative content analysis is designed systematically and transparently using Mayring's quality criteria. By observing criteria such as content validity, transparency, replicability and theoretical foundation, the scientific quality of the analysis can be ensured. The results are thus well-founded and comprehensible, making a significant contribution to answering the research question.

The multi-stage research design reflects a methodological triangulation approach that incorporates different perspectives and thereby has the potential to enhance the validity of the findings.

4.7.3 Potential for Bias

In assessing validity, particular attention must be paid to potential distortions arising from interest-driven statements. Since the expert interviews were conducted with representatives of various institutions, it can be assumed that their responses may be influenced by institutional perspectives and their own strategic interests. For example, it is plausible that a representative of a railway company evaluates the dynamics of competition with road freight transport differently than an actor from the trucking logistics sector. However, such differing assessments should not be viewed as a weakness, but rather as an indication of the heterogeneity of perspectives. They highlight that the analysis does not merely reflect the lowest common denominator, but instead acknowledges the diversity of viewpoints within the field under study, thereby enabling a more nuanced and differentiated understanding.

4.7.4 Optimization according to the 7S Model

The McKinsey 7S Model offers a systematic, holistic, and practice-oriented approach to the strategic analysis and development of organizations. Its key strength lies in its integrated consideration of both hard factors (e.g., strategy, structure, systems) and soft factors (e.g., values, skills, leadership style), providing a realistic and comprehensive foundation for managing organizational change. This multidimensional perspective makes the model particularly suitable for addressing complex organizational challenges. In the context of European freight transport, other strategic tools are also used to guide corporate direction, although these pursue more specific analytical objectives.

Porter's Five Forces Model is used to assess the competitive intensity within an industry by analyzing five structured forces: industry rivalry, the bargaining power of suppliers and buyers, the threat of new entrants, and the threat of substitute products. It enables the evaluation of industry attractiveness and supports strategic positioning.

The Business Model Canvas by Osterwalder is a visual tool for modeling and analyzing business models based on nine building blocks. It is especially applied in innovation-driven contexts, particularly within start-ups. The Balanced Scorecard by Kaplan and Norton links strategic objectives with measurable indicators across four perspectives: financial, customer, internal processes, and learning and growth. It serves as a tool for strategy implementation and performance measurement.

The ADKAR Model by Prosci focuses on individual change processes and addresses five core phases (Awareness, Desire, Knowledge, Ability, and Reinforcement), offering a people-centered perspective on managing change effectively. The TOWS Matrix is a methodological extension of the classic SWOT analysis and is used to derive concrete strategic options based on identified strengths, weaknesses, opportunities, and threats.

While these models are well-established in management research and widely applied in practice, they are not utilized within the scope of this academic study. This is because they either address different levels of analysis or serve specific purposes that do not align with the focus of this investigation. For the present analysis, the McKinsey 7S Model represents the most appropriate framework, as it provides the necessary organizational holism and enables a coherent integration of structure, culture, and strategy.

5 IMPLEMENTATION AND EVALUATION

5.1 Results of the arena analysis

In the following, the relevant stakeholders for the sustainability issues in European freight transport are identified and the associated employee survey topic areas are described. The stakeholders presented in Table 4, along with the relevant thematic areas from the employee survey, have been identified and elaborated specifically for the purposes of this dissertation. In an increasingly networked and complex economic world, various stakeholders with their specific employee survey topic areas play a central role. Companies themselves focus on securing their continued existence and developing new market segments through strategic expansion, promoting innovation and building strong customer relationships. Quality, sustainability and the expansion of IT are at the heart of their strategic considerations.

For supply chain managers, supply chain control and resource management are essential. They focus on optimizing value chains through targeted KPI management to increase efficiency and productivity within the company. IT managers as leaders in the IT sector, such as CIOs, are focused on driving IT security issues and future-oriented IT and AI solutions such as cloud computing, machine learning and data-driven decision-making platforms. IT providers complement these efforts by providing and operating relevant IT infrastructures and systems, replacing existing systems and implementing new market requirements.

Through analyses and surveys, market researchers provide sound insights into new trends and opportunities, which are indispensable for strategic decisions. The logistics sector, as a source of expertise for all areas of transport, ensures optimized transport times and cost reductions through process control and system integration. Infrastructure experts are responsible for the planning and modernization of facilities such as roads and loading infrastructure, thereby laying the foundation for efficient logistics and mobility. Examples of this include the expansion of multimodal transport hubs, the establishment of charging infrastructure for electric vehicles and the modernization of rail networks to overcome current challenges such as capacity bottlenecks and environmental regulations. Through its ministries, the state defines political framework conditions aimed at maintaining order and fostering sustainable development.

Associations such as the VDV or the Pro-Rail Alliance take on socio-political issues and regulatory tasks that are solved for companies at a higher level. Software manufacturers develop and implement innovative IT solutions that meet the digital requirements of companies. Finally, management boards, including MDs, CEOs and CFOs, are the colonel of decision-makers responsible for ensuring the economic stability and continued existence of their companies. Their strategic decisions have a lasting impact on the direction and growth of companies.

Freight companies with global operations across land, rail, and air transport sectors were included in the analysis. The group mainly comprises major firms such as DHL and Schenker, alongside single-mode transport providers including Swiss Rail and Deutsche Bahn. Small and medium-sized companies were not considered due to their lower market power and position on the market, as this large number of companies would have gone beyond the scope of this study. Consulting firms like Deloitte and CGI, with activities spanning the entire value chain, represent key players within this sector. CIOs of large freight operating companies were also identified as stakeholders and potential candidates for the expert interviews. Companies that offer software and infrastructure solutions for the entire industry are also important for the selection of relevant interview partners. It is also planned to interview ministries and transport associations, as these set the decisive guidelines and objectives, particularly the environmental objectives. Government representatives from the Ministry of Transportation and the European Commission are to be included as interviewees.

No shippers and suppliers were interviewed because these companies pursue a different objective and therefore also represent a different target group. Shippers and OEMs with their requirements and market position are not relevant for this work. Generally, what matters most to customers is the price-performance ratio. The terms and costs associated with the provision of this service frequently play a subordinate role. In addition, the scope would be far too extensive for qualitative research, especially as only the largest companies can be surveyed. Furthermore, each target group has its own requirements and there are too many niche markets that need to be considered.

To evaluate current trends, particularly in the areas of green energy and mobility, discussions were scheduled with current book authors, researchers and professors. The objective is to formulate optimization strategies in addition to anticipating the trends and changes on the horizon.

No response was received from the ministries and political bodies contacted, even after repeated inquiries. This implied that no discussion took place in this area. There was also no response to letters that were also created and sent as an alternative. Likewise, there was no response from the Chambers of Commerce and Greenpeace or from some university professors who have written relevant articles on freight transport, mobility and green energy. Potential experts were contacted several times to conduct the expert interviews, but no feedback was received. Numerous experts were not interviewed, and no personal exchanges took place.

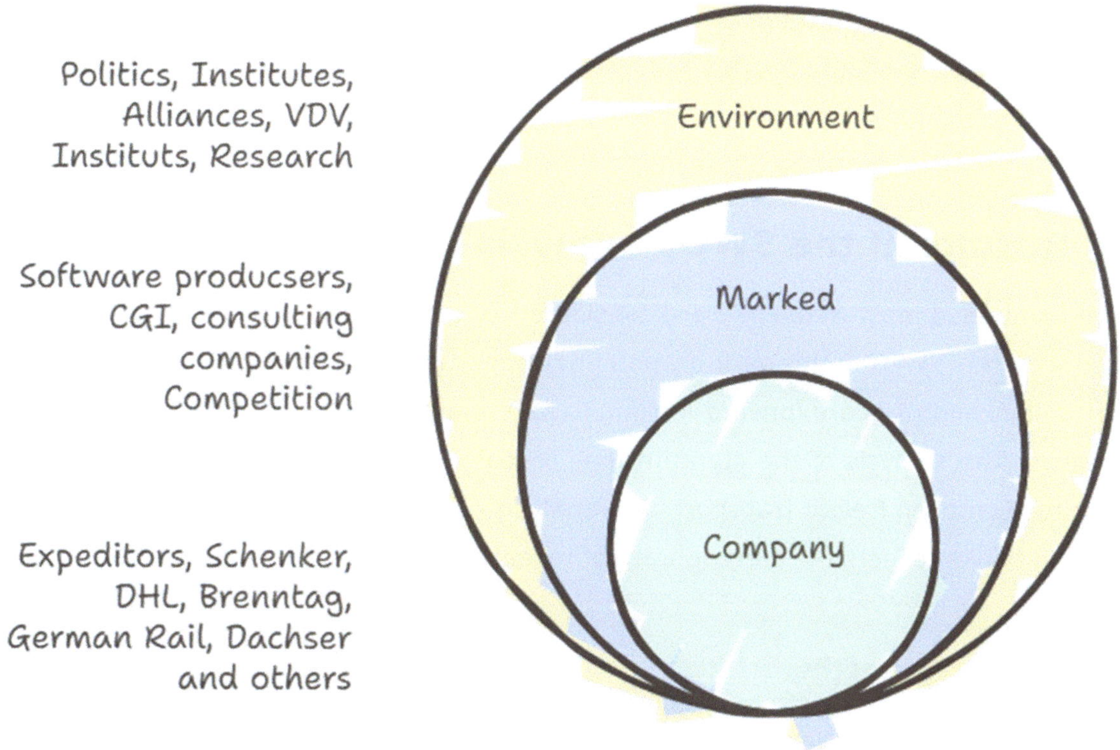

Figure 15: Application of the extended environment analysis
Source: Author' own illustration based on Pfaff 2020.

Each of these categories has specific characteristics and includes different actors that are relevant for understanding the dynamics of the company.

- **Company:** This level comprises the individual company and its internal stakeholders. These include employees, management, shareholders and other internal players who are directly involved in the business processes.
- **Market:** The market encompasses the company's immediate competitive environment, which includes competitors, customers, suppliers, and partners. Market trends, competitive strategies, customer needs and preferences as well as supplier relationships are examined in this area.
- **Environment:** The environment refers to the external, macroeconomic environment in which the company operates. This includes political, economic, social, technological, environmental and legal factors. At this level, the focus is on analyzing the wider framework conditions and trends shaping the business environment.

The Arena analysis is a useful tool in strategic management for systematically identifying and analyzing the various factors influencing a company. This approach aids in comprehending the operational contexts of transport companies and underpins informed strategic decisions. This structured approach enables participating companies to make well-founded strategic decisions by considering the diverse influences and interactions within their environment.

5.2 Results of the SWOT analysis

Topics mentioned most frequently are considered pivotal for the analysis and are integrated into the SWOT framework as potential opportunities and risks. The identified opportunities and risks are forward-looking in nature and were derived from the expert interviews. The analysis of the literature has revealed numerous strengths and weaknesses for European freight transport. The analysis of journals, specialist articles, websites, and books reflects the current state of research.

Strengths

- Transport companies are encouraged to work on their mobility concepts
- The Cargo Tube is an innovative approach to reducing CO2 emissions
- Depending on the length of the overhead line network, overhead line trucks can achieve economic advantages over long distances
- Greening as a Response to Noise Pollution
- Mobility and digitalization are just some of the issues being addressed in European transport policy
- Even today, rail is already the most climate-friendly means of transportation
- Electrification and hydrogen are gaining importance in replacing fossil fuels
- The future of freight transport is closely linked to the use of green energy

Weaknesses

- There is a growing concern among companies that their public image may convey a higher degree of environmental responsibility than is reflected in their internal operations
- At 31%, freight transport is the third largest economic sector that produces GHGs
- Road transport is still the leading mode of transport with diesel trucks
- The rail network must be further expanded so that cross-border traffic can continue to grow
- European road traffic is threatened by a shortage of drivers, permanent traffic jams and associated cost increases
- Only 60 percent of the German rail network is electrified, and even less in Europe

Figure 16: SWOT analysis - strengths and weaknesses
Source: Author's own representation.

The future of sustainable freight transport relies on innovative concepts and technologies aimed at reducing environmental impacts and increasing efficiency. Transport companies are encouraged to revise their mobility strategies to align with sustainability goals. Innovations such as the Cargo Tube present promising approaches to significantly reduce CO_2 emissions. For long-distance freight, overhead line trucks demonstrate economic advantages depending on the extent of the overhead line infrastructure. Additionally, initiatives focused on *greening against noise* aim to mitigate the environmental and social impacts of transportation systems. Mobility and digitalization remain central pillars of European transport policy, highlighting the integration of smart technologies to optimize logistics and reduce carbon footprints. Even today, rail transport remains the most climate-friendly mode of freight transportation.

Freight transport remains a significant contributor to greenhouse gas emissions, accounting for 31% of total emissions and ranking as the third-largest economic sector in this regard. Despite sustainability efforts, road transport, primarily reliant on diesel trucks, continues to dominate the sector. Companies face growing concerns about the discrepancy between their environmental image and actual practices, raising the issue of greenwashing. Meanwhile, the rail network, with only 60% electrification in Germany and even lower rates across Europe, requires substantial expansion to support cross-border traffic growth and improve transport efficiency. Additionally, European road transport is increasingly burdened by driver shortages, persistent congestion, and rising operational costs, all of which threaten its long-term viability and efficiency. Addressing these challenges necessitates a balanced approach combining infrastructure investments, decarbonization strategies, and transparent sustainability practices to ensure an environmentally and economically sustainable freight transport system.

The analysis of the literature has revealed numerous strengths and weaknesses for European freight transport. An analysis of academic and professional sources, including journals, articles, websites, and books, reveals the current state of af-fairs.

Based on the findings of the theory section and intensive study of literature, the author of this dissertation came to the conclusion as to which deductive categories are relevant for this work. These categories are described and derived in Table 15. The deductive categories, developed from the reviewed literature, provide the conceptual framework for the subsequent, interview-based formation of subjective categories.

Opportunities

- Promotion of rail freight transport reduces CO2 emissions
- Shifting freight traffic from road to rail
- Integration and use of new technologies such as digital automatic coupling digitizes freight
- Hybrid drives and the digitalization of transport processes increase efficiency and sustainability
- Automation of existing infrastructure and processes support mobility
- Alternative drive technologies such as hydrogen and e-fuels can reduce the ecological footprint of freight transport
- Reducing dependence on fossil fuels.
- Economic benefits through subsidies and progressive CO_2 taxation motivate people to invest in green technologies
- In addition to maintenance, AI also controls the traffic itself and monitors loading

Risks

- High consumption of resources for electrification
- The expansion of electrification requires large quantities of raw materials, the extraction of which can lead to considerable environmental damage
- Insufficient infrastructure for alternative drive systems
- Slow development of an infrastructure for hydrogen and e-fuels
- Lack of charging structures for electric drives is delaying the changeover
- High costs place an economic burden on companies
- Technological dependence on new technologies such as AI and innovative drives can lead to risks
- Social impact of autonomous systems could lead to job losses
- Social injustice and dissatisfaction among employees
- Hydrogen is not sufficiently available in terms of capacity

Figure 17: SWOT analysis - opportunities and risks
Source: Author's own representation.

A systematic comparison of the literature-based and expert-informed SWOT results was conducted to closely align theoretical perspectives with empirical findings. The analysis examined the extent to which the strengths and weaknesses identified in the academic literature were confirmed by the experts' statements, and which additional opportunities and threats were introduced through the interviews. In this way, the study provides a theory-driven yet empirically validated contribution to the assessment of the research field.

Code	Deductive categories	Explanation
C1	Potential and risks	Identifying potential and risks is essential, as many different perspectives are considered by the experts. In turn, measures for the future can be derived from the potential and risks.
C2	Ecological sustainability	The aim is to achieve the climate targets that have been set and reduce emissions as much as possible. Environmental dimensions are therefore of particular importance.
C3	Social sustainability	Social sustainability is an important aspect, as there are strong fluctuations in freight transport.
C4	Economic sustainability	Besides corporate image, economic considerations are paramount for companies, as they must maintain profitability to remain competitive.
C5	Technical sustainability	In 2025, digitalization and technical innovations are still in vogue and are still being developed. Technology is the component that has changed the most in recent years and will continue to change rapidly.
C6	Green Energy	The concepts of green and clean energy frequently emerge as key themes in efforts to optimize freight transportation. Various options and alternatives will be mentioned in the expert interviews.
C7	Further aspects	Another aspects is a collective term for all topics that the expert would like to mention or that have occurred to him/her. C7 is the last question (exit from the Interview).

Table 15: Derivation of the deductive categories
Source: Author's own representation.

The subjective categories are derived from the expert interviews. Other or additional categories can be formed for future and further research topics in this area. The deductive category system is described in detail in Appendix 19.

5.3 Results of the expert interviews

In this chapter, the statements from the expert interviews are described and examined based on individual questions.

Question B1 deals with identifying the greatest potential for increasing sustainability in freight transport. The majority of experts advocate for shifting freight transport from road to rail, as this transition promises significant efficiency gains. In particular, the consolidation of loads and the electrification of rail freight are emphasized as key areas of potential (cf. Experte 1 2024: Z12; cf. Experte 4 2024: 43). The digitalization and networking of transport users also offers great potential for increasing efficiency, although this potential is still underestimated by many market players (cf. Experte 5 2024: Z65). "An important technological advance in this area is digital automatic coupling (DAC), which could enable significant improvements in rail freight transport" (Experte 8 2024: Z96).

The attractiveness of rail transport must be increased to ensure sustainable success in the long term (cf. Experte 9 2024: Z34). Further potential is seen in piggyback transport, which provides for the integration of rail and road, particularly through automation and digitalization, which enable the piggyback transport of trucks over long distances (cf. Experte 8 2024: Z208). Energy generation and the short range of battery-powered vehicles also represent optimization potential (cf. Experte 13 2024: Z13; cf. Experte 11 2024: Z192). To reduce energy consumption, freight trains should run through without intermediate stops (cf. Experte 8 2024: Z105). In addition, it is emphasized that rail technology must not only serve climate and nature protection, but also noise protection, which could be achieved through optimized track layouts and wheelsets (cf. Experte 8 2024: Z144). The modernization of outdated railroad lines is also necessary to increase the attractiveness of rail (cf. Experte 11 2024: Z69; cf. Experte 2 2024: Z282). Sven Experte 11 further states: "Finally, the formation of new alliances is seen as an important factor in increasing efficiency in freight transport" (Experte 11 2024: Z55).

Question B2 examines the risks that can arise when increasing sustainability in freight transport. "According to experts, the electrification of freight transport entails considerable risks. In particular, the extraction of the required raw materials can lead to considerable environmental damage, and the availability of these materials may be limited" (Experte 6 2024: Z77). Therefore, it is recommended that the concept of sustainability be approached in a more comprehensive manner (cf. Experte 6 2024: Z78). In addition, electrification is not considered feasible everywhere, especially in urban and rural areas, where the costs are high compared to the benefits (cf. Experte 3 2024: Z107). Hydrogen is being discussed as an alternative technology, but its infrastructure also requires considerable expansion (cf. Experte 8 2024: Z133). Solar energy also poses a challenge, as the costs for its use in truck operations are considered prohibitively high (cf. Experte 1 2024: Z44). Furthermore, the limited range of battery-powered trucks is criticized, as these vehicles are currently unable to cover 1,000 kilometers without intermediate charging (cf. Experte 7 2024: Z126). Combined transport is also rated as inefficient and costly, which has led to a shift of loads back to the road (cf. Experte 8 2024: Z225; cf. Experte 3 2024: Z270).

Expert Experte 10 emphasizes: "The electrification of freight transport entails considerable risks. In particular, the extraction of the required raw materials can lead to considerable environmental damage, and the availability of these materials may be limited" (Experte 6 2024: Z77). In addition, the neglect of infrastructure, particularly the lack of a dedicated rail network for freight transport, is identified as a significant risk. These structural deficits and the slow progress in creating fair competitive conditions between modes of transport represent major challenges for the sustainable development of freight transport (cf. Experte 9 2024: Z53; cf. Experte 3 2024: Z116).

Question C1 examines how European freight transport can ensure environmental sustainability in the future. The experts emphasize rail as the most sustainable mode of transport, especially for the use of hydrogen compared to trucks (cf. Experte 1 2024: Z18). "According to Experte 3, in addition to emissions, the energy source for propulsion is also crucial" (Experte 3 2024: Z58). Hydrogen, biofuels and synthetic fuels are mentioned as alternatives in the event of insufficient electrification (cf. Experte 6 2024: Z38; cf. Experte 10 2024: Z59; cf. Experte 11 2024: Z197). A third of the experts are against internal combustion engines and prefer electric drives, although hydrogen and fuel cells are also considered promising (cf. Experte 12 2024: Z107; cf. Experte 7 2024: Z79). However, the increasing demand for electricity requires sufficient power availability and the expansion of overhead lines for trucks, despite the potential environmental impact (cf. Experte 3 2024: Z88). In addition, the expert Experte 1 is of the opinion that "more environmentally friendly alternatives such as induction or conductor rail lines are also being considered" (Experte 1 2024: Z86).

Optimum utilization of all modes of transport is crucial to avoid empty runs, and hybrid drives for rail and road are recommended (cf. Experte 3 2024: Z628). Measures such as the reduction of wind resistance through aerodynamic vehicle designs also contribute to ecological sustainability (cf. Experte 6 2024: Z95). The expansion of infrastructure, including new freight terminals and battery-powered traction units, is also considered necessary (cf. Experte 3 2024: Z146). It is emphasized that railroads should operate entirely on 100% renewable energy (cf. Experte 3 2024: Z601), whereas progress in fuel cell technology is viewed as restrained (cf. Experte 7 2024: Z78). Combined transport and the *last mile* remain key challenges (cf. Experte 3 2024: Z132, Z170).

The expert Experte 10 is of the opinion that "truck traffic also has sustainability potential and should be made as emission-free as possible" (Experte 10 2024: Z57). A holistic view of ecological and economic efficiency is essential (cf. Experte 10 2024: Z93). Innovative approaches such as the greening of buildings with moss to reduce CO_2 and beautify the city are also proposed (cf. Experte 12 2024: Z102). Overall, the experts see the reduction of diesel locomotives and the abandonment of internal combustion engines as key steps towards a sustainable future for freight transport in Europe (cf. Experte 12 2024: Z329).

Question C2 addresses the necessary improvements in the construction of new infrastructure as well as the development of novel propulsion technologies in freight transport. "We absolutely must minimize the carbon footprint when building new infrastructure," emphasizes Experte 5 (2024: Z36). A key issue is the environmentally friendly production and sufficient availability of green hydrogen, although its current supply is problematic (cf. Experte 6 2024: Z66; cf. Experte 7 2024: Z280). Synthetic fuels such as e-fuels could be used as an alternative, especially as the nationwide supply of hydrogen is not yet secured (cf. Experte 7 2024: Z282). Research and development in the field of fuel cells and e-fuels is crucial to advancing future drive technologies (cf. Experte 7 2024: Z80; cf. Experte 4 2024: Z123). A switch to electric drives in truck traffic is also recommended to reduce environmental pollution (cf. Experte 5 2024: Z83). Automation and digitalization are seen as important approaches to increasing ecological and economic efficiency. "We need to optimize both outdated freight cars and transshipment points in a targeted manner," explains Experte 8 (2024: Z154, Z171).

"Cargo tubes offer enormous potential for economic sustainability and could be laid underground in the future to significantly minimize environmental impact," explains Experte 2 (2024: Z50). However, the development and introduction of new drive technologies require a solid infrastructure (cf. Experte 6 2024: Z128), whereby alternative drive systems should be developed first before the corresponding infrastructure is added (cf. Experte 6 2024: Z115). Overall, minimizing the carbon footprint, developing sustainable drive systems and optimizing existing systems through automation and digitalization are crucial for the sustainable future of freight transport.

Question D1 explores which social dimensions are likely to gain prominence in the future of freight transport. The experts underscore the imperative for a profound paradigm shift among shippers, customers, and employees, particularly concerning the adoption of sustainable practices and heightened environmental responsibility (cf. Experte 3 2024: Z36). Measures are frequently endorsed only when they do not entail additional costs, whereas many market participants perceive the long-term return on investment (ROI) as insufficient and remain apprehensive about potential financial risks (cf. Experte 6 2024: Z31; cf. Experte 4 2024: Z77). In addition, a reduction in consumption by consumers could generally reduce production (cf. Corn 2024: Z88). Openness to technology is seen as essential to create a consensus on sustainable measures (cf. Experte 9 2024: Z94). Ethical principles and social standards must be adhered to, whereby greenwashing should be prevented and compliance with these standards should be made transparent (cf. Experte 12 2024: Z145; cf. Experte 11 2024: Z79).

There is an imbalance in working conditions and pay, particularly in the HGV sector. There are calls to create affordable overnight accommodation for drivers, avoid wage dumping and ensure fair working conditions (cf. Experte 1 2024: Z58; cf. Experte 3 2024: Z214). Foreign companies, in particular, frequently remunerate workers below the legally mandated minimum wage, a practice that necessitates urgent rectification (cf. Experte 3 2024: Z255). In rail transport, physically demanding tasks, such as coupling wagons, are increasingly perceived as unattractive, which is why automation could provide relief, but requires training (cf. Experte 8 2024: Z157; cf. Experte 5 2024: Z117). Process optimizations should not be at the expense of employees, and autonomous driving should not endanger jobs (cf. Experte 5 2024: Z121). Occupational safety, especially when handling hazardous substances, is another relevant aspect. Automation could eliminate stressful activities and create more pleasant workplaces (cf. Experte 12 2024: Z131; cf. Experte 8 2024: Z265). "We should reduce truck traffic and shift more freight traffic to rail in order to effectively reduce noise pollution", recommends Experte 7 (2024: Z192). Overall, the social aspects of sustainability must be given comprehensive consideration, including improving working conditions, promoting ethical standards and minimizing social impacts through new technologies (cf. Experte 10 2024: Z85).

Question D2 deals with measures and incentives to increase social acceptance of sustainability in freight transport. The experts emphasize the need for a comprehensive expansion of infrastructure, especially for HGVs, including additional truck stops, which requires considerable government investment (cf. Experte 3 2024: Z284). The introduction of digital automatic coupling (DAC) is considered beneficial for rail transport, as it both reduces the physical strain on employees and promotes digitalization and efficiency in freight transport (cf. Experte 8 2024: Z299).

"We need a general willingness to change processes and more transparency for everyone involved in order to effectively drive forward the expansion of infrastructure and the implementation of sustainable measures," demands Experte 9 (2024: Z262). The expansion and introduction of new technologies also require considerable investment (cf. Experte 3 2024: Z284). Tax breaks and subsidies could serve as incentives for environmentally friendly technologies, while green transportation could also be promoted by external providers (cf. Experte 6 2024: Z160). Progressive CO_2 taxation could also support the switch to sustainable transport solutions (cf. Experte 8 2024: Z294). Overall, the experts advocate for an integrated approach that combines targeted investments, process optimization, financial incentives, and regulatory interventions to enhance both social acceptance and operational efficiency in sustainable freight transport.

Question E1 examines how freight transport can achieve economic sustainability. The experts emphasize the importance of corporate image and the need for responsible use of resources. However, companies frequently hesitate to make investments due to concerns over potential reputational damage v(cf. Experte 5 2024: Z60, Z64). The sustainable use of resources is crucial for long-term economic sustainability (cf. Experte 3 2024: Z67). Combined transport is seen as the key to economic sustainability, as it makes efficient use of infrastructure and promotes cooperation between modes of transport. This contributes to increasing capacity and reducing CO_2 emissions but requires sufficient wagons and load carriers (cf. Experte 3 2024: Z261, Z318; cf. Experte 2 2024: Z175). Efficiency increases in freight transport also require optimized departure and braking controls as well as the promotion of new drive technologies such as hydrogen (cf. Experte 8 2024: Z163; cf. Experte 6 2024: Z235).

The expansion of rail networks and transshipment points is essential for ecological and economic sustainability. The deliberate selection of materials in the planning of railway stations significantly contributes to both sustainability and architectural appeal (cf. Experte 12 2024: Z93-98). Financing models, such as tax concessions and increased tolls could reduce unnecessary journeys while sharing concepts and the earmarked use of vehicle tax offer alternative approaches (cf. Experte 2 2024: Z158; cf. Experte 1 2024: Z105; cf. Experte 10 2024: Z200). Strategic alliances play a pivotal role in enhancing transport connectivity, particularly with regard to the 'last mile', which continues to pose a significant challenge (cf. Experte 11 2024: Z106-110; cf. Experte 12 2024: Z192). The experts emphasize the central importance of logistics as the backbone of society, as functioning logistics are essential for prosperity and stability (cf. Experte 4 2024: Z203). A sustainable organization of freight transport is therefore indispensable (cf. Experte 12 2024: Z72).

Question E2 examines which economic measures can be taken to relieve the burden on freight transport infrastructure in the future. A key strategy for alleviating pressure on infrastructure lies in the more conscious and sustainable use of natural resources (cf. Experte 3 2024: Z45). Nevertheless, societal backing for sustainability initiatives often wanes when these are associated with elevated financial burdens (cf. Experte 2 2024: Z149). Measures for cost efficiency and optimization are therefore necessary, such as the promotion of digital interlocking, which can improve route utilization (cf. Experte 3 2024: Z410). The expansion and connection of the infrastructure must be consistently driven forward, whereby transparency about the service life of the infrastructure is necessary for well-founded investment decisions (cf. Experte 12 2024: Z66; cf. Experte 10 2024: Z107).

Alliances and freight optimization can significantly contribute to relieving congestion on transport routes, as can the avoidance of long-term construction sites (cf. Experte 11 2024: Z110, Z131). The outdated infrastructure must be modernized to meet current requirements, whereby automation technologies can further increase efficiency (cf. Experte 1 2024: Z23; cf. Experte 8 2024: Z444). Improving the efficiency of road infrastructure requires the optimization of truck freight operations and a more effective use of transport space, for example through the integration of high-capacity vehicles like gigaliners and the expansion of hub-and-spoke logistics networks (cf. Experte 6 2024: Z208). Constant running speeds could also improve the flow of traffic and reduce energy consumption (cf. Experte 8 2024: Z448). The financing of infrastructure projects may be enhanced through variable, usage-based charges, aimed at promoting a more efficient allocation of traffic loads (cf. Experte 9 2024: Z187). A combination of digital technologies, optimized capacity utilization, infrastructure modernization and targeted regulation through financial incentives is necessary to sustainably relieve the burden on freight transport infrastructure.

Question F1 examines how technology in freight transport could change in the coming years. Most of the experts forecast significant technical developments, particularly in the areas of digitalization and networking (cf. Experte 2 2024: Z281). Digitalization and automation are seen as key success factors, including the development of new assistance systems to increase efficiency and reduce workload (cf. Experte 3 2024: Z491). The expert Experte 9 is also of the opinion that "One focus of technical progress lies in the improvement of vehicle technology, especially hybrid locomotives and trucks, as well as the conversion of diesel locomotives to hybrid drives" (Experte 9 2024: Z266).Fuel cells and optimized battery technologies are also considered promising, with a particular need to reduce battery costs and charging times (cf. Experte 7 2024: Z181). Lighter wagons made from new composite materials and improved aerodynamics could further increase efficiency (cf. Experte 8 2024: Z419; cf. Experte 4 2024: Z55).

The use of 3D printing is seen as having the potential to significantly reduce the need for transportation (cf. Experte 8 2024: Z311). The harmonization of European rail networks and the introduction of IT and AI to control traffic flows and avoid empty runs could further increase efficiency (cf. Experte 11 2024: Z159; cf. Experte 9 2024: Z229). The future technical development of freight transport will be shaped by digitalization, automation, new materials, 3D printing and AI. These technologies aim to increase efficiency and promote economic sustainability.

Question F2 deals with how information technology and AI can contribute to increasing sustainability in freight transport. Experte 4 emphasizes: "IT and AI are seen as central instruments for optimizing freight transport, as they accelerate the exchange of information and improve cooperation between players and countries" (Experte 4 2024: Z52).Especially the optimization of the travel chain through the use of AI, for example for the evaluation of camera images for load monitoring and damage detection, contributes to efficiency and safety (cf. Experte 12 2024: Z240-242). AI can also digitize and accelerate processes in maintenance depots, reducing maintenance costs and increasing the availability of vehicles (cf. Experte 12 2024: Z267).

The expansion of infrastructure remains a critical priority, with IT and AI offering the potential to significantly enhance the efficiency of planning and implementation processes (cf. Experte 3 2024: Z101). AI can help to shorten construction times and monitor environmental changes to ensure sustainability (cf. Experte 12 2024: Z255, Z260). The increase in efficiency through AI is also supported by better control of traffic flows and minimization of empty runs (cf. Experte 8 2024: Z326). Experte 5 further states: "Although the use of AI requires investment, the potential benefits outweigh the costs, as they can significantly improve the economic and ecological efficiency of freight transport" (Experte 5 2024: Z218).

Question G1 examines what incentives policymakers could create to promote the use of green energy in freight transport. Effective political steering mechanisms constitute a key lever in expediting the transition toward sustainable energy systems. Current approval processes in Europe are often too lengthy and can take up to seven years (cf. Experte 8 2024: Z427). Streamlining these processes could encourage investment in green technologies. Subsidies are seen as another measure to motivate companies to use green energy and increase rail transport (cf. Experte 1 2024: Z151). In addition, higher taxation of fossil fuels and the introduction of CO_2 taxes could contribute to the promotion of sustainable energy (cf. Experte 1 2024: Z154; cf. Experte 9 2024: Z246).

The expansion of accessible infrastructure for alternative fuels is crucial, as is supporting banks in financing green technologies and closing grid gaps through electrification (cf.

Experte 2 2024: Z249). Reducing subsidies for fossil fuels could accelerate the transition to sustainable energies (cf. Experte 7 2024: Z236). A balanced energy mix could serve as a transitional solution until a complete switch to green energy is possible (cf. Experte 6 2024: Z231). Promoting competition in the infrastructure sector is essential for fostering innovation and enhancing efficiency (Experte 9 2024: Z177). Political incentives such as the streamlining of approval processes, subsidies, CO_2 levies, infrastructure investments and the promotion of fair competition could accelerate the transition to green energy in freight transport and increase sustainability.

Question G2 addresses the essential transformations in infrastructure, technology, and propulsion systems required to reduce greenhouse gas emissions for the benefit of future generations. The experts emphasize the importance of legal regulations and political decisions for the transition to more sustainable solutions. The legally enshrined coal phase-out in Germany illustrates the effectiveness of regulatory frameworks (cf. Experte 5 2024: Z225). In addition to regulatory measures, sustained and targeted communication with shippers, customers, and the broader public is essential. Experte 2 elaborates: "Creating transparency and raising awareness of the need for sustainable change" (Experte 2 2024: Z256).

The expert Experte 7 clarifies "The promotion of carbon-free drive technologies is seen as a decisive competitive advantage, whereby e-fuels can only be used to a limited extent in long-distance transport due to the required energy density" (Experte 7 2024: Z250, Z261). Comparative analyses of carbon-intensive and renewable electricity generation are necessary to uncover efficiency potentials and inform targeted strategies for greenhouse gas mitigation (cf. Experte 6 2024: Z157). The expansion of the charging infrastructure for battery-powered HGVs in combined transport is essential to promote low-emission drives (cf. Experte 3 2024: Z617). The experts call for a coordinated strategy of political measures to achieve a significant reduction in greenhouse gases in freight transport.

Question H1 solicits further feedback from experts or highlights other issues that they consider significant regarding the sustainable development of freight transport. The experts emphasize the need for simplified cross-border cooperation in order to improve efficiency and coordination in freight transport (cf. Experte 2 2024: Z279). "This also includes shortening the time-consuming approval processes for new products", Experte 2 continues (Experte 2 2024: Z71). Many companies are hesitant due to a lack of business cases and the perceived costs of sustainable changes (cf. Experte 7 2024: Z99), although an accelerated development towards sustainability is necessary to achieve the climate targets.

Furthermore, experts highlight the necessity of establishing specialized freight transport infrastructure tailored to the sector's unique requirements, including measures for the environmental enhancement of rail facilities to increase sustainability. (cf. Experte 8 2024: Z390). According to expert Experte 4, "there is a consensus that more engineering services will be required over the next 30 years in order to implement technical and infrastructural changes" (Experte 4 2024: Z153). The results of the expert interviews show that there is broad agreement among the participants on key issues. In many areas, a shared understanding emerges, especially regarding the importance and direction of sustainability goals.

However, some differences in opinion are also evident, particularly concerning the role of the state. While some experts argue in favor of stronger government regulation, others believe that the state should not take on direct control functions but should instead provide greater financial support to promote sustainable development. No significant disagreements were observed among participants, which could be attributed to shared support for sustainability goals.

From a research perspective, different stakeholder groups place emphasis on varying aspects. Representatives of associations, for instance, underline the need for stronger collaboration and the expansion of consortia. Companies, by contrast, frequently act in accordance with strategic self-interests, placing a stronger emphasis on technological innovation as a means of consolidating or enhancing their market position.

5.4 Review of the quality criteria

Mayring's quality criteria are relevant for research based on interpretative approaches. A review of the quality criteria comes to the following conclusions:

1. procedural documentation: To ensure the traceability of the research work, the individual steps of data collection must be comprehensively documented. A detailed description of the research process enables other scientists to understand and evaluate the analysis. The research process was described in Chapter 4.

2. argumentative support for the interpretation: The interpretation of the results must be supported by arguments. It should be clear why a particular interpretation is considered plausible. This requires the interpretations to be underpinned by quotes. To avoid misunderstandings, all interpretations of the expert interviews were supported by quotes with exact details of the expert and the line number.

3. rule-based: Qualitative research should be conducted according to clearly defined rules. Qualitative content analysis according to Mayring is based on a rule-guided procedure that aims to minimize interpretation errors. The central systematic rules include transparency and comprehensibility, the minimization of interpretation errors and the standardization of the method used.

4. proximity to the object: The proximity of the researcher to the object of investigation is an essential element of qualitative research. Mayring emphasizes that the results must always be interpreted within the context of the material under study. The interpretation should be closely oriented to the statements of the interviewees. Proximity to the object of investigation is ensured, as the researcher is actively engaged in the field of European freight transport and possesses a profound understanding of its thematic challenges. The experts interviewed are leading specialists in the freight transport and transport logistics sectors. The topic and the subject of the dissertation were discussed in detail with all of the experts in advance, which means that the interpretation is closely aligned with the statements of the interviewees.

5. communicative validation: To increase the validity of the interpretations, these should be checked through feedback from the respondents. This includes submitting the results for confirmation to ensure that the participants have been accurately represented. Transcripts were sent to respondents for review and confirmation. In two cases, adjustments were made to clarify linguistic misunderstandings due to dialects. All interviewees confirmed the accuracy and approved the transcripts.

6. triangulation: The use of multiple methods, perspectives or researchers helps to increase the validity and reliability of the research results. Triangulation ensures that the research is viewed from different perspectives, which strengthens the objectivity and reliability of the results. Several methods and perspectives were considered. Several researchers have accompanied the methodological process and provided important impulses for the research approaches. The work and its results were analyzed from different perspectives to ensure the greatest possible objectivity.

The quality criteria established by Mayring, designed to ensure that qualitative research is systematic, transparent, and verifiable, were thoroughly applied and successfully met. Consequently, these criteria are considered to have been fulfilled in the present study.

5.5 Derivation of optimization recommendations

A SWOT analysis was created from the results of the literature analysis and the expert interviews conducted. Three specific recommendations are generated, which are derived from McKinsey's 7S model (cf. Consultport 2024).

No.	Factor	Element	Description
1	Strategy	Hard	The strategy of European freight transport is to optimize mobility and create sustainability. The colonel's goal is to achieve the climate targets that have been set. This includes the creation of new drive systems, the promotion of electrification and the refurbishment of technical facilities.
2	Structure	Hard	European freight transport involves transport companies that transport shipments on all modes of transport on behalf of customers. Given the size of the European freight operating company, one might assume that it is managed by a strict hierarchical order. However, this is not the case.
3	Systems	Hard	European freight transportation is one of the most important cornerstones of the economy and has some of the most efficient systems in the world. These systems serve all IT tasks from marketing to operations in the supply chain. The company is also constantly innovating to improve its existing systems.
4	Shared values	Soft	Core values of the industry following megatrends such as mobility and environmental protection. The industry is undergoing a major changing social norms and values due to electric mobility and the associated issues.
5	Skills	Soft	Employee skills must emphasize qualifications and training in new technical topics to ensure that employees are able to work economically and efficiently.
6	Style	Soft	Road transport employees are poorly paid and have a heavy workload. The sector is characterized by high fluctuation.
7	Staff	Soft	Critic economic consideration involves shifting the focus toward employees and their workplace conditions. Strategies to improve employee retention must be intensified, especially as staff are progressively influenced by the adoption of new systems and technological progress.

Table 16: Recommendation for action in the 7S model according to McKinsey
Source: Author's own illustration based on McKinsey.

This results in the following specific recommendations:

- Shift from road to rail
- Electrification and promotion of renewable energy sources
- Promoting technical developments such as AI and innovation topics

5.6 Interim conclusion

The dissertation examines the relevant actors and employee survey topic areas in European freight transport that contribute to the promotion of sustainability. The central stakeholders are freight operating companies, IT providers, infrastructure managers, government bodies, associations and consultants. Companies such as DHL, Schenker and Deutsche Bahn play a decisive role due to their market power, while IT providers and managers increase efficiency and reduce emissions through digitalization and automation. Consulting firms such as Deloitte and CGI provide strategic support in the field of sustainability.

Political players, including the European Commission, are creating the regulatory framework and setting environmental targets. Shifting traffic from road to rail is seen as promising, as is the electrification and digitalization of rail freight transport. Technologies such as DAC can significantly improve the efficiency of rail transport. Combined transport, i.e. the integration of rail and road, is seen as another model for reducing emissions.

The electrification of freight transport harbors risks, for example due to the environmental impact of mining raw materials for batteries. Hydrogen is an alternative, but there are infrastructural hurdles. It is emphasized that an equivalent infrastructure is required for all modes of transport. Challenges also exist in the cost and time issues of combined transport, with a return to road transport being perceived as a risk.

Ecological sustainability in freight transport can be achieved through better utilization of transport modes, the use of low-emission technologies and the reduction of empty runs. Greening urban infrastructure and reforestation measures by companies are also emphasized. The electrification of terminals and the use of hybrid drives also play a central role. The carbon footprint should be minimized in the construction of new infrastructure. The promotion of rail transport as an alternative to air freight, alongside the development of cargo tubes and emerging propulsion technologies (e.g., hydrogen and e-fuels), constitutes a crucial component of sustainability efforts.

The social aspects of sustainability include improving the working conditions and remuneration of truck drivers and ensuring ethical standards. Autonomous driving and digitalization have the potential to enhance working conditions; however, they necessitate comprehensive training. Political incentives such as tax breaks and subsidies for green technologies as well as CO_2 taxation could promote the use of sustainable energy. The expansion of overhead lines and the creation of a nationwide infrastructure for alternative fuels are crucial.

Technological innovations such as 3D printing, hydrogen and e-fuels technologies as well as digitalization and automation will have a significant impact on the future of freight transport. The optimization of battery technologies and the use of AI for traffic control and route planning are key elements for sustainable development.

This dissertation illustrates that the transformation of European freight transport towards sustainability is a multidimensional process that requires close cooperation between stakeholders, technological innovation and political support. The development of infrastructure, alternative drive systems and technology will be crucial to make freight transport ecologically and economically sustainable and to achieve the European climate targets. A SWOT analysis highlights both the opportunities and risks involved in implementing sustainable strategies. Three specific recommendations were formulated based on the McKinsey 7S framework.

The transition to a sustainable freight transport system relies on a modal shift from road to rail, leveraging rail's higher energy efficiency and lower emissions. Simultaneously, electrification and the use of renewable energy sources are essential to decarbonize both rail and road transport. Advancements in technological developments, including artificial intelligence and other innovations, play a pivotal role in optimizing logistics processes and improving operational efficiency. Collectively, these measures constitute the foundation for a resilient and future-oriented transport infrastructure.

6 CONCLUSION

The results of this dissertation are summarized and critically discussed in this chapter. In addition to the assumptions made, the results are also critically scrutinized. The forward-looking section offers an outlook on the need for further research and summarizes the proposed optimization measures. The conclusion provides a critical appraisal, thoroughly examining the limitations of the study.

6.1 Answering the research questions

This chapter provides answers to the individual research questions and implications for science. The previous research results are primarily used as a basis for information. The following answers thus represent a compression and classification of the qualitative evaluation carried out in the context of the research question.

What are the potential benefits and risks associated with competition in the mobility sector?

In the mobility competition, there is potential to increase efficiency, particularly by shifting freight transport from road to rail. This step enables significant efficiency gains, as consolidated loads and the electrification of rail freight transport make key contributions to more sustainable logistics. In addition, the digitalization and networking of transport users offers enormous potential for further optimization, an aspect that is still underestimated by many market players. One important technological advance in this area is DAC, which enables significant improvements in rail freight transport. This technology as well as targeted automation and digitalization could increase the attractiveness of rail transport and secure it in the long term. Another starting point is piggyback transport, which involves the integration of rail and road. Here, piggyback transportation of trucks over long distances offers considerable potential for efficiency gains.

There is also potential for optimizing energy generation and range of battery electric vehicles. To further reduce energy consumption in rail transport, freight trains should run through without intermediate stops. Modernized rail technology not only contributes to climate and nature conservation but can also improve noise protection. The modernization of outdated railway lines is essential to further enhance the attractiveness of rail transport. In addition, the formation of new alliances between the various players in freight transport is an important factor in increasing efficiency and making mobility competition sustainable.

In the mobility competition, the electrification of freight transport entails considerable risks. The extraction of the required raw materials can lead to considerable environmental damage and the availability of these materials is limited. These factors make it necessary to take a more comprehensive view of the sustainability concept. In addition, electrification is not considered to be feasible across the board, as the high investment costs are often not in proportion to the benefits. Hydrogen is being discussed as an alternative technology, but its infrastructure also requires considerable expansion. The use of solar energy for trucks is also seen as problematic, as the costs for its use are prohibitively high. Added to this is the limited range of battery-powered trucks, which further limits their efficiency. Combined transport, widely regarded as a sustainable solution, is often perceived as less competitive due to inefficient and costly processes, which may incentivize a modal shift back to road freight.

Energy policy is also a decisive factor, with the right electricity mix being seen as central to success. The purchase of CO_2 certificates, on the other hand, is seen as an inadequate solution. The neglect of infrastructure, particularly the lack of a dedicated rail network for freight transport, poses a significant risk. These structural deficits and the slow progress in creating fair competitive conditions between modes of transport are major challenges for the sustainable development of freight transport.

To what extent will European freight transport have to be adapted to achieve the sustainability goals by 2040?

Meeting the sustainability objectives for European freight transport by 2040 necessitates extensive transformations at the technological, infrastructural, and organizational levels. These adjustments affect all stakeholders and require a holistic view as well as close cooperation between politics, industry and society.

A central goal must be the electrification of freight transport. This includes the reduction of diesel locomotives and the switch to electric mobility as well as the use of hybrid drives where full electrification is not possible. The expansion of overhead lines and the introduction of alternative technologies, such as routes via induction or over the ground, are crucial here. In addition, batteries, drives and overhead lines must be made more sustainable. As the range of battery-powered vehicles is limited, the availability of green hydrogen must also be ensured on a large scale, and pipelines must be built to supply it. The use of synthetic fuels could be a transitional solution, but as their use is limited, the focus must be on carbon-free technologies in the long term.

European freight transport is outdated in many respects and needs to be modernized. This applies to both the rail and road infrastructure. The modernization and expansion of transshipment points, the digitalization of signal boxes and the automation of processes are key measures. At the same time, new terminals need to be built to make logistics more efficient. The expansion of charging infrastructures for battery-powered trucks and the improvement of the hydrogen infrastructure are further necessary steps. Network expansion is also necessary, incorporating innovative concepts such as cargo loops and hyperloops, which may potentially be implemented underground in the future.

Increased efficiency in freight transport can be achieved by reducing unnecessary journeys and promoting a sensible mix of rail and road. Maximizing the load capacity of transport vehicles leads to improved utilization rates, yielding both economic and environmental benefits. The introduction of digital automatic coupling (DAC) can significantly improve shunting efficiency. Optimizations in transport chains, such as direct connections without reloading and the creation of a common rail network in Europe, are crucial to achieving sustainability goals.

In addition to technological and infrastructural changes, social and ethical aspects are also important. A rethink among customers and employees is required to promote sustainable logistics processes. Compliance with minimum wages and ethical principles, the avoidance of greenwashing and the creation of transparency and traceability must be enforced. Improving working conditions in freight transport is also essential, for instance by reducing strenuous manual labor and increasing the number of visual display unit (VDU) workstations.

To achieve the sustainability goals, the carbon footprint of freight transportation must be drastically reduced. This requires a comprehensive switch to renewable energy, the promotion of biofuels and the electrification of transport routes. Forests must be reforested, not cleared, and surfaces should not be sealed. Planting vegetation on buildings and noise control measures are also important steps towards reducing environmental pollution. The energy mix plays a central role here, and the use of green electricity and hydrogen must be prioritized.

Government support is essential to promote the necessary adjustments in freight transport. This includes the creation of a legal framework, the introduction of subsidies and incentives, for example through CO_2 levies or the taxation of fossil fuels. Improved cooperation between countries is also necessary to build a harmonized European rail network. Permanent construction sites should be avoided, and infrastructure measures should be designed transparently, including a clear presentation of the lifespan of infrastructure.

Continuous research and development of new technologies, especially in the areas of battery and hydrogen technologies, is essential to facilitate the transition toward sustainable mobility solutions. In addition, work must be done on new, lighter transport containers and wagons, which enable a higher payload due to their lower dead weight and thus increasing efficiency.

To achieve the sustainability targets by 2040, European freight transport must make extensive technological, infrastructural and social adjustments. Increased electrification, the expansion of hydrogen infrastructure, the digitalization and automation of processes and the creation of a sustainable energy mix are of central importance. In addition, close cooperation between all stakeholders is required to ensure environmentally friendly and efficient logistics.

What specific measures can help to make European freight transport more sustainable in line with the climate targets that have been set?

The transition to more sustainable freight transport necessitates comprehensive measures involving transport companies, policymakers, society, and technological advancements. The following discussion outlines key areas of action encompassing ecological, economic, and social dimensions.

Measures taken by transport companies: To reduce the environmental impact of freight transport, companies must increasingly shift transportation from road to rail. This not only helps to reduce CO_2 emissions but also reduces traffic congestion and noise pollution. A decisive step is the electrification of driving systems in both rail and road transport. This should be driven using renewable energy, especially in terminal operations where a switch to sustainable energy sources is possible. The production of transport services must be CO_2-neutral. This can be supported by the increased use of synthetic fuels such as biofuels. A key element of the decarbonization strategy is the reduction of diesel locomotives in favor of electric mobility, alongside the electrification of road transport. To further minimize the consumption of resources, companies should avoid unnecessary journeys and empty runs. It is crucial to push forward the switch from fossil fuels to renewable energy sources and to modernize the fleet by converting vehicles to hybrid technologies.

The formation of alliances and associations can improve the utilization of transport capacities and strengthen the market position of companies at the same time. In addition, processes within the transport chains should be made more efficient through optimized logistics planning. Another measure to reduce transport volumes is the establishment of 3D printing centers, whereby only basic materials need to be transported instead of finished products.

Measures by all market participants: Shifting transportation to rail not only reduces CO_2 emissions but also contributes to noise reduction. In addition to the rail infrastructure, trucks should also be made quieter and transshipment points should be equipped with noise control measures, such as noise barriers and planted buildings. The greening of transshipment sites also contributes to CO_2 reduction, as plants absorb CO_2 and produce oxygen. Transshipment sites should also be converted so that they can use green hydrogen. All market participants need to be networked to develop a holistic view of the market and jointly drive forward sustainability efforts.

Technical measures: Technological change in freight transport requires the increased use of ecological technologies. Hybrid drives should be further developed and made more efficient. Simultaneously, intensive research into alternative fuels is essential, as carbon-based energy sources currently possess the highest energy density necessary for transporting heavy loads. Further technical optimization concerns the weight reduction of vehicles and loading units through the use of new composite materials, which enable a higher payload due to a lower dead weight. In addition, outdated wagons and locomotives must be brought up to the latest state of the art, and electrification of the rail network should be intensified to reduce emissions and noise pollution.

Political measures: Politicians must take action to create a uniform basis for freight transport in Europe. This especially means enhancing international cooperation and extending the European rail network. The concept of *platooning* for trucks should also be introduced in order to reduce energy consumption by reducing the distances between vehicles. However, a corresponding legal basis is required for this. Other policy measures include the reforestation of deforested areas, the promotion of green drive technologies and the expansion of overhead lines to promote the electrification of transport. In addition, legal measures, such as the coal phase-out and the modernization of existing transport infrastructure, should be implemented without delay. Eco-certifications and the definition of carbon footprints could also contribute to sustainability in freight transport.

Society must adapt to the new developments in freight transport and develop an awareness of how to better use natural resources. A holistic view of the entire supply chain is of great importance here. In addition, more engineering work will be needed in the coming decades to ensure the further development of driven technologies and energy sources. Particular attention should be paid to the reduction of certificate purchases and a balanced carbon footprint.

In the area of social sustainability, it is of central importance to take a critical look at the use of subcontractors to avoid wage dumping. The minimum wage must be guaranteed regardless of the origin of the personnel, and compliance with ethical principles and fair working conditions should be enforced. Social justice must also be ensured throughout the sector. Other measures include protecting employees from hazardous substances and cleaning products and creating more VDU workstations to improve working conditions.

AI offers numerous opportunities to make freight transport more efficient. For instance, artificial intelligence can be employed to advance driver assistance systems and optimize digital interlocking technologies. AI can also be used to monitor traffic data, terrain and infrastructure to shorten construction times and minimize the distance between trains. The use of AI can also optimize maintenance measures and increase efficiency in production.

To ensure a transparent and sustainable infrastructure, the service life and condition of the existing transport infrastructure must be openly presented. Permanent construction sites over long periods of time must be avoided in order not to impair the efficiency of freight transport.

In the area of alternative drive systems, fuel cells, e-fuels and battery-powered drive systems could play an important role. It is necessary to carefully select suitable alternatives and possibly combine different energy sources to achieve comprehensive decarbonization. Hydrogen is also a promising alternative that should be explored further. The transformation of freight transport requires a comprehensive and multidimensional approach that combines technical, environmental, political and social measures. A sustainable and efficient future for European freight transport can only be achieved through the integrated implementation of these interrelated areas of action.

How can green energy be used sensibly in European freight transport to achieve sustainability goals?

The sensible use of green energy in European freight transport faces several risks and challenges that need to be addressed. A central challenge lies in the long development cycles required to bring products to market, which significantly hinders progress. Subsidies can serve as an incentive to foster innovation and encourage investment in green technologies. The production of energy by companies themselves could also be promoted, either through direct subsidies or by taxing fossil fuels to increase the incentive to switch to green alternatives.

Another hurdle is the availability and affordability of the necessary infrastructure. Banks should actively support the financing of green projects and gaps in the grid must be closed by expanding the infrastructure. In addition, a CO_2 levy could serve as an incentive to switch to sustainable energy. Close cooperation with policymakers is crucial to promote the reduction of fossil fuels and diversify the energy mix. Legal regulations, such as the coal phase-out, also play a central role. It is also essential to communicate the benefits of sustainability to shippers, customers and the public to raise awareness of the need for a sustainable transformation. Sustainability must be made transparent and certificate purchases must be reduced to achieve long-term environmental goals. Selfish market interests should be minimized to promote collective goals.

An additional risk concerns the challenge of selecting suitable alternative energy suppliers, which is crucial for ensuring reliability, cost-efficiency, and sustainability. Charging infrastructures for batteries must be expanded and the promotion of green drive technologies must be stepped up to drive forward the electrification of freight transport. Although carbon-based energy sources have the necessary energy density, e-fuels can only be used to a limited extent. The comparison between "dirty" and "green" electricity must be continuously considered to maximize the benefits for companies. Cooperation among European countries must be enhanced, particularly with regard to supply chains and the ship-side import infrastructure for hydrogen. The entire supply chain must be considered to increase efficiency. Finally, society needs to adapt to market trends and support a shift towards more sustainable solutions.

6.2 Limitation of work

Every doctoral project is subject to limitations in scientific research. One potential challenge in this project is the selection of suitable experts. The search for experts proved to be difficult. There was a risk that the optimal experts might not be selected. Alternatives, such as focusing the research project on German freight transport or restricting it to specific modes of transport, were deliberately ruled out by the author due to the overarching importance of the topic.

The study is explicitly limited to freight transport within Europe. Public passenger transport and private journeys by car or other means of transport are not considered in this study. There is also no discussion of sustainability and mobility issues on other continents, and no comparison is made with European freight transport. Various methodological approaches were considered to collect the relevant topics and questions for the expert interviews.

Another methodological limitation is the focus on specific modes of transport. The focus of this study is on freight transport via road, rail, and air; inland waterway and intermodal maritime transport are deliberately excluded from the scope of analysis. In this context, digital aspects of sustainability are also not specifically addressed.

An additional limitation was the absence of some of the planned experts. To mitigate the risk of participant dropout, a deliberate over-recruitment of experts was carried out. Nevertheless, some experts did not respond despite written requests or were unable to be interviewed due to capacity bottlenecks. Notably, no expert interviews were conducted with political representatives.

A uniform methodological framework was created for all experts by conducting interviews using a standardized interview guide. The interviews took place remotely via video calls. Due to licensing and security restrictions for some experts who were unable to use Zoom, the interviews were conducted via Microsoft Teams instead.

The category system developed in this study represents a further potential limitation. Although the categories that were meaningful and relevant for the study were covered, a single study cannot cover the entire thematic spectrum. Future research could develop further categories on this basis and expand the employee survey topic area.

6.3 Implication for science

This work closes an important research gap by adding a technical perspective to the classic three-pillar model of sustainability (ecological, economic, social). This fourth pillar is especially significant in relation to the processes of automation and digitalization. In addition to these aspects, technical sustainability also includes technological innovations and expansions that enable the modernization of European freight transport. The proposed causal model makes it clear that sustainable development in the transport sector remains incomplete without taking technical factors into account. Future research should further investigate and expand this technical aspect. The political framework conditions in the European context in particular offer potential for further work. A key focus could be the harmonization of rail transport and the creation of a legal basis for progressive CO_2 taxation. An evaluation and reorganization of freight forwarding law could also help to promote technical sustainability. The focus should be on discussing political objectives and their concrete implementation.

In addition to the political dimension, the application of IT and AI solutions in futurology could be further intensified. Artificial intelligence offers a wide range of possibilities that have frequently been underestimated in practical implementation to date. Numerous innovations and trends in this area, particularly regarding automation and data analysis, require in-depth investigation and implementation. Although the work thoroughly examines the technical dimension, social factors were necessarily only partially explored. In rail transport, social issues are largely regulated, whereas in road transport there is a considerable need for action. Future research could focus on reducing labor turnover and employee motivation measures to promote sustainable human resource strategies in the transport sector.

Another important research topic is the economic dimension of technical sustainability. Here it would be useful to investigate the costs and operating conditions of alternative drive technologies such as hydrogen, batteries and fuel cells. Research should be carried out into the economic and ecological conditions under which the use of these technologies is profitable and sustainable. This dissertation primarily examined large companies and leading players in European freight transport. However, future work could focus more on small and medium-sized companies to better understand how they deal with sustainability and mobility. International comparisons are also possible, for example between Bulgaria and the USA or China, as well as between Europe, Asia and North America, to analyze country-specific challenges and solutions.

Finally, the sector of courier, express, and parcel services (CEP) constitutes a previously underexplored field of research. CEP services, which involve the transportation of low-weight and low-volume general cargo, play an increasingly important role in European freight transport and should also be considered in future studies. By integrating technical sustainability, this work offers significant scientific added value by demonstrating that the modernization of European freight transport is necessary not only to achieve climate targets but also to ensure long-term competitiveness and infrastructure.

6.4 Optimization recommendation for freight transport

The literature analysis and expert interviews revealed various relevant perspectives on the future development of freight transport. These findings were explored in greater depth in chapter 5.3 by means of a SWOT analysis and based on this, recommendations were formulated using McKinsey's 7S model. These recommendations are further detailed and supplemented in this chapter.

1. shift from road to rail transport

Shifting freight transportation from road to rail is a key measure for reducing CO_2 emissions and noise. Despite years of effort, combined transport has not yet established itself due to long transit times and high costs. Strengthening rail transport requires political, technical and financial measures. The rail infrastructure must be improved, and transshipment points must be equipped with noise control measures such as noise barriers and green buildings, which will also help to reduce CO_2 emissions. The establishment of a decision-making body at European level could help to monitor and coordinate progress.

2. electrification of transport and promotion of renewable energies

Electrifying drive systems in rail and road transport constitutes an essential measure for advancing sustainability goals. This requires the expansion of renewable energies and the promotion of efficient drive technologies. To support research and development in this area, government funding must be made available and research institutions must receive increased support. The EU and national governments could create incentives and set up corresponding funding programs.

3. modernization of vehicle fleets and use of alternative drive technologies

To accelerate the transition to low-emission mobility, it is necessary to convert vehicle fleets to alternative drive systems, such as hybrid technologies or synthetic fuels. However, this conversion is associated with high costs that individual companies are often unable to bear. Government subsidies and support programs are needed to drive this transformation forward.

4. use of green hydrogen and strategic alliances

Transforming transshipment points and terminals to operate on green hydrogen represents an important step forward in achieving low-emission logistics systems. Closer networking of all market participants is necessary to develop a holistic market perspective and drive forward sustainable innovations. This could be facilitated by strategic alliances or transit authorities, which particularly enable smaller companies to gain easier market access.

5. technological optimization and digitalization

Technical optimizations, such as reducing the weight of vehicles by using lighter composite materials, contribute to increasing efficiency. Outdated wagons and locomotives need to be modernized, with the introduction of digital automatic coupling playing a central role. This technology not only reduces the physical workload of staff but also enables digitalization and the use of sensors on freight cars. AI can play a decisive role here in data processing, monitoring and control.

6. political measures and legal framework

Political measures are needed to improve cooperation between European countries and develop a common rail network. In particular, the introduction of technologies such as *platooning* in truck transport, which are already established in countries such as Canada, could be promoted through appropriate legislative changes. Other measures include the reforestation of deforested areas, the promotion of green drive technologies and the expansion of overhead lines for the electrification of transport.

7. sustainability certificates and social justice

The introduction of eco-certifications and the definition of carbon footprints for transport companies could also contribute to sustainability in freight transport. Companies that do not comply with these standards should expect sanctions such as the non-approval of loading equipment or the refusal of credit. Special attention must also be paid to social justice in the sector. Fair working conditions, regardless of the origin of the employees, should be ensured through training and regular inspections. Companies should also take measures to retain their employees in the long term and avoid staff turnover.

8. role of artificial intelligence

AI plays a central role in the optimization of freight transport. In addition to monitoring functions, it offers potential for driver assistance systems and the control of traffic data. By using AI, maintenance processes can be optimized, construction times shortened, and resources used more efficiently. The continuous expansion of high-performance IT systems and internet connections is essential here.

In summary, the transformation of European freight transport requires a multidimensional approach that combines technical, environmental, political and social aspects. Only through the coordinated interaction of these measures can a sustainable and efficient future of freight transport be ensured.

6.5 Outlook

The scientific outlook of this dissertation focus on the development and analysis of a holistic and interdisciplinary model to promote sustainability in freight transport. The focus is on ecological, economic, technological as well as social and political fields of action. The change in freight transport towards more sustainability requires a systemic approach that integrates several dimensions.

A central starting point of this dissertation could be the development of an integrative model for the decarbonization of freight transport, which includes the shift of transport from road to rail, the electrification of drive technologies and the increased use of renewable energies. This could involve investigating how different measures interact with each other and the extent to which synergies exist between technological and logistical optimizations and political and social measures.

The investigation of technical developments, such as the electrification of drive systems, increasing efficiency through alternative fuels and the use of hybrid and hydrogen-based drive systems, could form a significant part of the research. It would be relevant to evaluate the potential, challenges and costs of these technologies in practice and to identify possible solutions for their implementation. One focus could be the further development of AI-based systems to make freight transport more efficient and safer. The role of political actors in the creation of a uniform European framework for sustainable freight transport would be another key research topic. Specific political measures could be analyzed here, such as the expansion of rail infrastructure, the promotion of overhead line technologies for trucks or the legal establishment of CO_2 balances and eco-certifications. International cooperation between European countries and the influence of global markets and regulations could also be considered here.

In addition to the technological and political aspects, the societal and social impact of change in freight transport should be highlighted. This includes society's changing awareness of resource consumption, the change in working conditions due to technological innovations, such as increasing digitalization and automation, and the ethical challenges of employing subcontractors. A key research focus should be on exploring the mechanisms to guarantee social acceptance and social justice in the development and implementation of sustainable transport systems.

Since the transition to sustainable freight transport is not limited to technological innovations or political measures alone, but requires the networking of various disciplines, the dissertation should take an interdisciplinary approach. This could be achieved through cooperation between engineering, logistics, economics and social sciences to gain a comprehensive understanding of the necessary changes. It would be important to develop scenarios and simulations that show how different measures reinforce or inhibit each other.

In the long term, this dissertation should provide a framework for further research, as it identifies open questions in the field of sustainable drive technologies, political regulation and social transformation. This could include the need for further empirical studies or simulations aimed at identifying the optimal combination of measures to make the most emission-free freight transport possible.

In summary, the result of the dissertation is to create a scientific basis for a sustainable transformation of freight transport that goes beyond technical innovations and includes ecological, economic, political and social dimensions. Thus, the dissertation contributes to the development of a low-emission and resource-efficient European freight transport system.

Literature

A

Accenture. 2024. Klimaeffekte der Digitalisierung 2.0, Studie zur Abschätzung des Beitrags digitaler Technologien zum Klimaschutz in Deutschland. bitkom. Berlin.

Adams, E. K., & Ellis, A. D. 2019. Green freight transportation and its impact on supply chain efficiency. *Transportation Research Part D: Transport and Environment*. Retrieved March 09, 2025, from https://doi.org/10.1016/j.trd.2019.09.018.

Alekseev, P. V. 2024. Assessment of the modern climate policy of the European Union. *Review of Business and Economics Studies*. Retrieved March 15, 2025, from https://rbes.fa.ru/jour/article/view/355.

Asana. 2025. *SWOT Analysis: Examples and Templates*. Retrieved May 01, 2025, from: https://asana.com/resources/swot-analysis.

B

Bagwan, N., et al. 2023. Kältetechnologien in gekühlten Lieferketten: Über die Eignung und Nachhaltigkeit von Trockeneis als Kühlmedium. In U. Clausen & M. Dellbrügge (Publisher), Interdisziplinären Konferenz für Produktion, Logistik und Verkehr (S. 45-58). Springer Publishing.

Baloise. 2024. What do we mean by mobility? Retrieved July 19, 2024, from: https://www.baloise.com/en/home/news-stories/news/all-news/2024/what-do-we-mean-by-mobility.html.

Banister, D. 2022. Unsustainable Transport: City Transport in the New Century (2nd ed.). Routledge.

Banister, D. 2023. Sustainable Transport: Planning for Walking and Cycling in Urban Environments (3rd ed.). Woodhead Publishing.

BDE. 2023. Europäische Kommission muss 2026 über die mögliche Einbeziehung der thermischen Abfallverwertung in das EU-ETS ab 2028 entscheiden. Retrieved November 18, 2024, from: https://www.bde.de/presse/europaeisches-emissionshandelssystem-europaeisches-parlament-stimmt-zu/.

Behrends, S., & Floden, J. 2022. Transport and Logistics Planning. Emerald Publishing.

Black, W. R. 2023. The Green Logistics Handbook: Sustainability and Environmental Impact. Springer Publishing.

Blauwens, G., De Baere, P., & Van de Voorde, E. 2022. Transport Economics (4th ed.). De Boeck Supérieur.

BMDV. 2023. Die Zukunft braucht Wege. Retrieved October 07, 2024, from: https://www.bmv.de/SharedDocs/DE/Artikel/G/der-infrastrukturdialog.html

BMWK. 2024a. Renewable Energy. Retrieved December 04, 2024, from: https://www.bmwk.de/Redaktion/DE/Artikel/Aussenwirtschaft/exportinitiative-energie.html.

BMWK. 2024b. ELEKTRA - Wasserstoff für Wasserwege. Redakteur: Ramon Henk. Retrieved December 19, 2024, from: https://www.bmwk.de/Redaktion/DE/Wasserstoff/Praxisbeispiele/ELEKTRA.html.

Experte 2, R. 2024. Experte 2.

Bothe, J. 2024. Die Energiewende im Alltagsverstand: Diskurse, Orientierungsmuster und Verdrängungsinstrumente. Universität München. Retrieved October 12, 2024, from: https://edoc.ub.uni-muenchen.de/33898/.

Bräkling, E., et al. 2020. Logistikmanagement: Mit Logistik-Power schnell, schlank und fehlerfrei liefern (2. Aufl.). Springer Gabler Publishing.

Brecha, R., & Nature, S. 2023. Decarbonising road freight transport: The role of zero emission technologies. Nature. Retrieved October 27, 2024, from: https://www.nature.com/articles/s41586-020-2796-5.

Browne, M. 2023a. Decarbonizing Transportation Systems. Springer Publishing.

Browne, M. 2023b. Urban Freight and Sustainable Cities: Integrating Sustainability into Urban Freight Transport. Routledge.

Brunner, M., et al. 2020. Impact of emerging transport technologies on freight economic and environmental performance: A system dynamics view. International Journal of Environmental Research and Public Health, 17(93092). Retrieved August 22, 2024, from: https://www.mdpi.com/1660-4601/17/9/3092.

Brüsemeister, T. 2008. Qualitative Forschung: Ein Überblick (2. Aufl.). Beltz Publishing.

BSL Transportation Consultants. 2020. Der Bericht 2020 über den kombinierten Verkehr. Retrieved October 14, 2024, from: https://bsl-transportation.com/uic-2020-report-on-combined-transport-presse-28-oktober-in-bruessel-vorgestellt/.

Business.gov 2025. Zero-emission zones to be introduced in many cities from 2025. Retrieved August 17, 2024, from: https://business.gov.nl/sustainable-business/sustainable-business-operations/zero-emission-zones-to-be-introduced-in-many-cities-from-2025/.

BVL. 2024. Ziele der Logistik. Retrieved August 29, 2024, from: https://www.bvl.de/service/zahlen-daten-fakten/logistikbereiche/logistik#:~:text=Allgemein%20gesehen%20bsteht%20%20die%20fortw%C3%A4hrende,und%20Kontrolle%20aller%20dazugeh%C3%%20B6rigen%20Aufgaben.

C

Caldeirinha, V., Felicio, J., & Pinho, T. 2023. Role of cargo owner in logistic chain sustainability. Sustainability. Retrieved November 02, 2024, from: https://doi.org/10.3390/su151310018.

Carter, P. N. 2023. Climate Change and Maritime Transport: Challenges and Opportunities. Springer Publishing.

CISL. 2025. Cambridge Institute for Sustainability Leadership. Comment: Sustainability is under unprecedented attack. Here's how we can fight back. Reuters. Retrieved May 22, 2024, from: https://www.reuters.com/sustainability/climate-energy/comment-sustainability-is-under-unprecedented-attack-heres-how-we-can-fight-back-2025-02-10/.

Climate Portal. 2024. Renewable Energy. Retrieved November 21, 2024, from: https://climate.mit.edu/explainers/renewable-energy.

CM Logistics. 2024. Güterverkehr: Bahn versus LKW. Retrieved November 21, 2024, from: https://cm-log.eu/gueterverkehr-bahn-vs-lkw/.

Cole, M. H. 2023. Eco-Friendly Transport and Logistics: Environmental Impacts of Transportation. Routledge.

Consultport. 2024. Das 7S Modell nach McKinesy. Retrieved November 22, 2024, from: https://consultport.com/de/erfolg-als-berater/das-7s-modell-von-mckinsey-anhand-eines-praktischen-beispiels-erklaert.

Coyle, J. J., Novack, R. A., Gibson, B. J., & Bardi, E. J. 2023. Transportation: A Supply Chain Perspective (11th ed.). Cengage Learning.

Crainic, T. G., & Kim, K.-H. 2023. Intermodal Transportation (3rd ed.). Edward Elgar Publishing.

Crawford, N. 2024. The green transition and European industry. *Survival, 66*(1), 99-124. Retrieved December 17, 2024, from: https://www.tandfonline.com/doi/full/10.1080/00396338.2024.2309078.

CT4EU Campaign. 2022. Durch die intelligente Vernetzung von Schiene, Straße und Wasserwegen trägt der intermodale Güterverkehr zu einer nachhaltigen Reduktion der Treibhausgasemissionen um bis zu 90 % und einer Verringerung des Energiebedarfs um bis zu 70 % bei. Retrieved August 23, 2024, from: https://www.ct4eu.eu.

Cullinane, S. 2024. Innovations in Green Logistics. Palgrave Macmillan.

Curtis, C. 2024. Handbook of Sustainable Transport: An International Perspective. Edward Elgar Publishing.

D

Dabija, D., & Nastase, C. 2024. Exploring the evolution of sustainability paradigms: From linear to circular economy models. Proceedings of the International Conference on Business Excellence, 18(1), 754–769. Retrieved December 29, 2024, from: 10.2478/picbe-2024-0066.

Dalrada. 2024. Climate Technology, Grüne Energie, Saubere Energie. Retrieved October 09, 2024, from: https://dalradaclimatetech.com/the-difference-between-green-energy-and-clean-energy/.

Davis, J., Thomas, J., Antony, S. V., Johnson, S. K., & Joseph, R. B. 2024. Smart automated solution for public transport system. *2024 10th International Conference on Advanced Computing and Communication Systems (ICACCS)*, 1, 1322–1327. Retrieved January 12, 2025, from: https://www.researchgate.net/publication/385198669_Smart_Automated_Solution_for_Public_Transport_System.

De Feo, G., et al. 2019. Economic, energy and environmental efficiency of road freight transportation sector in the EU. Energies, 12(3874). Retrieved August 22, 2024, from: https://doi.org/10.3390/en12203874.

Deineko, E., et al. 2023. Sondierende Analyse des Verkehrssystems für das modulare Fahrzeugkonzept Operated als Personen- und Güterverkehr unter Verwendung von Kontinuumsannäherungstechniken. In U. Clausen & M. Dellbrügge (Publisher), Interdisziplinären Konferenz für Produktion, Logistik und Verkehr (S. 75-89). Springer Publishing.

Dellbrügge, M., Clausen, U., et al. 2023. Widerstandsfähigkeit von städtischen Fabriken und ihren Lieferketten. In U. Clausen & M. Dellbrügge (Publisher), Interdisziplinären Konferenz für Produktion, Logistik und Verkehr (S. 103-119). Springer Publishing.

Denteh, E., Danyo, A., Asamoah, J. K., Kyem, B. A., Addai, T., & Aboah, A. 2025. Integrating travel behavior forecasting and generative modeling for predicting future urban mobility and spatial transformations. arXiv preprint arXiv:2503.21158. Retrieved April 16, 2025, from: https://arxiv.org/abs/2503.21158.

Experte 11, S. 2024, Expert 11.

Deutsche Bahn. 2024a. Aktuelle Berichte des DB-Konzerns und der DB AG. Retrieved November 27, 2024, from: https://ir.deutschebahn.com/de/berichte/db-konzern-und-db-ag/.

Deutsche Bahn. 2024b. Nachhaltigkeit, Wasserstoff für die Schiene, Wasserstoff im Testbetrieb. Retrieved November 27, 2024, from: https://nachhaltigkeit.deutschebahn.com/de/massnahmen/wasserstoff/h2goesrail.

Deutsche Bundesstiftung Umwelt. 2024. Grundlagen einer nachhaltigen Entwicklung. Retrieved November 27, 2024, from: https://opac.dbu.de/ab/DBU-Abschlussbericht-AZ-30564-Studienbrief1.pdf.

Digital Experts. 2024. Megatrend Digitalisierung. Retrieved January 04, 2025, from: https://www.digital-experts.com/blog/megatrend-digitalisierung-trends-und-treiber-des-digitalen-wandels-und-die-frage-nach-dem-superlativ#:~:text=Laut%20der%20IDC%2DStudie%20%E2%80%9EData,so%20die%20Herausgeber%20 der%20Studie.

Dincer, H., & Yüksel, D. 2024. Economic Development and the Environmental Ecosystem: The Role of Energy Policy in Economic Growth. Springer Publishing.

Diwo, M. 2024. Trends und Megatrends der Digitalisierung: Die heimlichen Treiber der digitalen Transformation. Retrieved March 21, 2025, from: https://technavigator.de/digitalisierung/trends/.

Domagala, J., & Kadlubek, M. 2023. Economic, energy and environmental efficiency of road freight transportation sector in the EU. Energies, 16(10461). Retrieved November 23, 2024, from: https://doi.org/10.3390/en16010461.

Dorfmann, H. 2022. Europäische Verkehrspolitik. In Mobilitäts- und Transportrecht in Europa (S. 213-230). Springer Publishing.

Dresing, T., & Pehl, T. 2018. Praxisbuch für Interview, Transkription und Analyse: Regelsysteme für die qualitative Forschung (8. Aufl.). Own Publishing.

Duin, H., et al. 2023. Die Planung von Hyperloop-basierten Cargo-Tubes-Routen für nachhaltige logistische Lösungen. In U. Clausen & M. Dellbrügge (Publisher). Interdisziplinären Konferenz für Produktion, Logistik und Verkehr (S. 133-145). Springer.

DVZ. 2023. Zuwachs der Gütermengen. Retrieved Juni 30, 2025, from: https://www.dvz.de/unternehmen/luft/detail/news/mengenzuwachs-in-2024-und-2025-erwartet.html.

E

Ebert, R., & Wu, H. 2023. Dynamisches Container-Routing-Problem in einem schienengestützten Hub-and-Spoke-Netz. In U. Clausen & M. Dellbrügge (Publisher). Interdisziplinären Konferenz für Produktion, Logistik und Verkehr (S. 157-170). Springer Publishing.

Eisenkopf, A. 2023. Mobilität auf dem Klimaprüfstand: Bahnverkehr. Wirtschaftsdienst. Retrieved June 04, 2024, from: https://doi.org/10.1007/s10273-022-3170-z.

EIZ Rostock. 2022. Europe direkt - Europa oder europäische Union (2022). Retrieved Juli 14, 2024, from: https://www.eiz-rostock.de/europa-oder-europaeische-union-was-den-unterschied-macht/.

Elbasha, A. M., & Abdellatif, M. M. 2025. A IoT-based smart traffic management system. arXiv preprint arXiv:2502.02821. Retrieved August 07, 2024, from: https://arxiv.org/pdf/2502.028211.

Ember. 2025. Global electricity review 2025. Retrieved August 19, 2024, from: https://ember-energy.org/latest-insights/global-electricity-review-2025/.

e-Mobile. 2024. Klimaschutz im Güterverkehr. Retrieved August 22, 2024, from: https://www.e-mobilbw.de/themen/klimaschutz-verkehr/gueterverkehr-logistik.

Energate. 2024. Tests von Wasserstoffzügen. Redakteur: Carsten Kloth. Retrieved August 17, 2024, from: https://www.energate-messenger.ch/news/207409/deutsche-bahn-testet-ab-2024-wasserstoffzug.

Energy News. 2024. Siemens sieht Wasserstoff als Zukunftsmarkt. Retrieved June 27, 2024, from: https://energynewsmagazine.at/2020/11/23/siemens-sieht-wasserstoff-zuge-als-zukunftsmarkt/.

Enste, D. & Potthoff, J. 2021. Behavioral economics and climate protection: Better regulation and green nudges for more sustainability. Retrieved June 28, 2024, from: https://www.econstor.eu/bitstream/10419/248810/1/1785225669.pdf.

Ernest & Young. 2024. Fortschrittsmonitor Energiewende 2024, BDEW und EY präsentieren aktuelle Kennzahlen zum Stand der Energiewende und Prognosen für anstehende Investitionen. Retrieved September 30, 2024, from: https://www.bdew.de/energie/fortschrittsmonitor-energiewende-2024/.

ETF. 2018. European Transport Workers' Federation, Studien zur sozialen.nd wirtschaftlichen Bedeutung des kombinierten Verkehrs für die Arbeitnehmer in der Transportbranche. Retrieved June 28, 2024, from: https://www.etf-europe.org.

Euramco. 2024. Green Energy - Energy fonds. Retrieved June 28, 2024, from: https://www.euramco-asset.de/glossar/green-energy-fonds/#:~:text =G%C3%BCtesiegel%20f%C3%BCr%20gr%C3%BCne%20Energie,wenig%20damit%20zu%20tun %20haben.

Euronews. 2025. The European Tube: Inside the project to launch a continent-wide rail system by 2040. Retrieved October 19, 2024, from: https://www.euronews.com/travel/2025/03/24/the-european-tube-inside-the-project-to-launch-a-continent-wide-rail-system-by-2040.

European Commission. 2022. Der europäische Green Deal. Retrieved August 02, 2024, from: https://ec.europa.eu/info/strategy/priorities-2019-2024/european-green-deal_de.

European Commission. 2023. Die EU fördert den kombinierten Verkehr durch die Combined Transport Directive (92/106/EEC). Retrieved January 12, 2025, from: https://ec.europa.eu/transport/modes/combined-transport_en.

European Commission. 2024. Neues Ziel für 2040: Empfehlung zur Verwirklichung der Klimaneutralität bis 2050. Retrieved January 11, 2025, from: https://climate.ec.europa.eu/eu-action/climate-strategies-targets/2050-long-term-strategy_de.

European Commission. 2024a. Die europäische Union in aller Kürze. Retrieved August 07, 2024, from:https://op.europa.eu/webpub/com/eu-what-it-is/de/.

European Commission. 2024b. Ein Schritt zur Klimaneutralität: Das neue Klimaziel der EU für 2040. Brüssel, Belgien: Europäische Kommission Veröffentlichungen. Retrieved August 07, 2024, from: https://climate.ec.europa.eu/eu-action/climate-strategies-targets/2030-climate-targets_de?prefLang=de.

European Commission. 2024c. Europäischer Green Deal. Retrieved August 22, 2024, from: https://commission.europa.eu/strategy-and-policy/priorities-2019-2024/european-green-deal_de.

European Commission. 2025. Green Deal Industrial Plan: Plugging the skills gap. Retrieved February 23, 2025, from: https://year-of-skills.europa.eu/green-deal-industrial-plan-plugging-skills-gap_en.

European Committee for Standardization (CEN). 2019. Standardisierungsinitiativen, die den kombinierten Verkehr effizienter machen sollen. Retrieved November 13, 2025, from: https://www.cencenelec.eu.

European Council. 2024. Klimaneutralität. Retrieved September 19, 2024, from: https://www.consilium.europa.eu/de/topics/climate-neutrality/.

European Environment Agency (EEA). 2024. Die EEA betont, dass der Kombinierte Verkehr ein wesentliches Instrument zur Erreichung der EU-Klimaziele ist. Retrieved November 16, 2024, from: https://www.eea.europa.eu/en/analysis/publications/european-climate-risk-assessment.

European Environment Agency. 2024. Klimaneutralität in Europa: Chancen und Herausforderungen. Kopenhagen, Dänemark: Berichte der EEA. Retrieved August 23, 2024, from: https://www.eea.europa.eu/en/analysis/publications/addressing-the-environmental-and-climate-footprint-of-buildings/addressing-the-environmental-and-climate-footprint-of-buildings.

European Parliament. 2016. Die Überprüfung der Combined Transport Directive 92/106/EEC im Jahr 2023 betont die Notwendigkeit, die Effizienz und Effektivität der Richtlinie zu verbessern. Retrieved October 24, 2024, from: https://www.europarl.europa.eu/doceo/document/TA-8-2016-0381_EN.html.

European Parliament. 2023. Green deal, key to a climate neutral and sustainable EU. Retrieved August 17, 2024, from: https://www.europarl.europa.eu/topics/en/article/20200618STO81513/green-deal-key-to-a-climate-neutral-and-sustainable-eu.

European Parliament. 2023. Schwedische Güterverkehrssysteme. Retrieved August 28, 2024, from: https://www.europarl.europa.eu/doceo/document/CRE-5-2002-09-24_DE.html?redirect.

European Parliament. 2024. Treibhausgasemissionen. Retrieved September 18, 2024, from: https://www.europarl.europa.eu/topics/de/article/20180301STO98928/treibhausgasemissionen-nach-landern-und-sektoren-infografik.

European Union. 2012. Vertrag über die Arbeitsweise der Europäischen Union (Vor dem Inkrafttreten des Reformvertrags von Lissabon, am 26.10.2012).

EWE. 2024. Was ist grüne Energie. Retrieved October 07, 2024, from: https://www.ewe.com/de/zukunft-gestalten/klimaschutz.

F

Federal Government. 2024. EU-Klimaschutzpaket Fit for 55. Retrieved August 10, 2024, from: https://www.bundesregierung.de/breg-de/service/archiv/fit-for-55-eu-1942402.

Federal Ministry for Digital and Transport. 2025. *Verkehrsprognose 2024/2025*. Retrieved June 17, 2024, from: https://www.bmv.de/SharedDocs/DE/Anlage/G/verkehr-in-zahlen24-25-pdf.pdf?__blob=publicationFile

Federal Statistical Office. 2021. Güterverkehr in der EU: Keine Verlagerung auf Bahn und Schiff. Retrieved July 30, 2024, from: https://www.destatis.de/Europa/DE/Thema/Verkehr/gueterverkehr-eisenbahn.html.

Federal Statistical Office. 2022. Güterleistung von Eisenbahn und Binnenschifffahrt, Pressemitteilung 148. Retrieved July 30, 2024, from: https://www.destatis.de/DE/Presse/Pressemitteilungen/2022/04/PD22_148_463.html?templateQueryString=verkehr.

Federal Statistical Office. 2024a. Beförderungsmengen des Güterverkehrs in Europa. Retrieved July 30, 2024, from: https://www.destatis.de/DE/Themen/Branchen-Unternehmen/Transport-Verkehr/Gueterverkehr/Tabellen/gueterbefoerderung-lr.html.

Federal Statistical Office. 2024b. Der Güterverkehr nach Verkehrsträgern sortiert. Retrieved July 30, 2024, from: https://www.destatis.de/Europa/DE/Thema/GreenDeal/_Grafik/11-gueterverkehr-anteil.html?nn=217548.

Federal Statistical Office. 2024c. Europäischer Green Deal, Sicherstellung der Klimaneutralität bis 2050. Retrieved August 02, 2024, from:https://www.destatis.de/Europa/DE/Thema/GreenDeal/_inhalt.html#798700.

Federal Statistical Office. 2024d. Menge und Leistung der Beförderungen. Retrieved August 02, 2024, from:https://www.destatis.de/DE/Themen/Branchen-Unternehmen/Transport-Verkehr/Gueterverkehr/Tabellen/gueterbefoerderung-lr.html.

Federal Statistical Office. 2025. Güterverkehr – Beförderungsmenge und Beförderungsleistung nach Verkehrsträgern. Retrieved May 25, 2025, from: https://www.destatis.de/DE/Themen/Branchen-Unternehmen/Transport-Verkehr/Gueterverkehr/Tabellen/gueterbefoerderung-lr.html.

Fedoskin, V., Bakulina, G. N., Chivileva, I. V., Martynushkin, A. B., & Polyakov, M. V. 2024. Formation of a system of indicators for assessing the availability and results of using freight transport. *E3S Web of Conferences*. Retrieved October 08, 2024, from: https://doi.org/10.1051/e3sconf/202451504016.

Financial Times. 2024. EU pushes for tech transfers from Chinese clean tech firms. Retrieved December 27, 2024, from: https://www.ft.com/content/f4fd3ccb-ebc4-4aae-9832-25497df559c8.

First Climate. 2024. Grünstrom für Unternehmen, von First Climate. Retrieved September 16, 2024, from: https://www.firstclimate.com/gruenstrom-naturstrom-komplettpakete.

Fischer, A. 2021. Klimaschutzstrategien im deutschen Verkehrssektor. Universität Stuttgart. Retrieved September 12, 2024, from: https://elib.uni-stuttgart.de/handle/11682/10800.

Fischer, A. 2024. Innovationen für eine grüne Logistik: Technologien und Prozesse. Springer Vieweg.

Fit for 55. 2022. Der Plan der EU für einen grünen Übergang – Consilium. Retrieved September 17, 2024, from: https://www.consilium.europa.eu/de/policies/green-deal/fit-for-55-the-eu-plan-for-a-green-transition/.

Fleischmann, B., & Klose, A. 2021. Quantitative Approaches in Logistics and Supply Chain Management. Springer Publishing.

Fockenbrock, D. 2020. Test des Wasserstoffzugs. Redakteur: Dieter Fockenbrock, Handelsblatt. Retrieved Dezember 14, 2024, from: https://www.handelsblatt.com/unternehmen/handel-konsumgueter/emissionsfreie-antriebe-deutsche-bahn-will-2024-wasserstoffzug-von-siemens-testen/26650498.html.

Franhofer ISE. 2024a. Wege zu einem klimaneutralen Energiesystem. Retrieved September 28, 2024, from: https://www.ise.fraunhofer.de/de/veroeffentlichungen/studien/wege-zu-einem-klimaneutralen-energiesystem.html.

Fraunhofer IFAM. 2024. Klimafreundlicher Wasserstoff und Brennstoffzellen. Retrieved September 28, 2024, from: https://www.ifam.fraunhofer.de/de/magazin/klimafreundlicher-wasserstoff-und-brennstoffzellen.html.

Fraunhofer ISE. 2024b. 55% bis 2030 und 100% bis 2050. Was bedeutet das CO2-Reduktionsziel der EU für die deutsche Energiewende? Retrieved September 27, 2024, from: https://www.ise.fraunhofer.de/de/presse-und-medien/presseinformationen/2020/55-prozent-bis-2030-und-100-prozent-bis-2050-was-bedeutet-das-co2-reduktionsziel-der-er-fuer-die-deutsche-energiewende.html.

Fraunhofer. 2024. Mobilitätskonzepte für Unternehmen. Retrieved Oktober 19, 2024, from: https://www.muse.iao.fraunhofer.de/de/leistungsspektrum/mobilitaetskonzepte_unternehmen.html.

G

Gatta, V., et al. 2019. Sustainable urban freight transport adopting public transport-based crowdshipping for B2C deliveries. European Transport Research Review. Retrieved August 04, 2024, from: https://doi.org/10.1186/s12544-019-0352-x.

Gattorna, J. 2019. Logistics 4.0: Digital Transformation of Supply Chain Management. Kogan Page.

Gavanas, N. 2024. New mobility services and sustainable urban development: The common EU vision and the challenges for transport planning in Greek cities. *ICTPW2024. Book of Proceedings*. Retrieved August 04, 2024, from: https://doi.org/10.37199/c41000710.

German association for foreign Business. 2024. Auf dem Weg zu einer nachhaltigeren Weltwirtschaft. Retrieved November 26, 2024, from: https://dgap.org/de/forschung/publikationen/auf-dem-weg-zu-einer-nachhaltigeren-weltwirtschaft.

German Institute for Economic Research. 2024. Berlin. Energie und Klimapolitik: Mehr Tempo für die Transformation. Retrieved December 09, 2024, from: https://www.diw.de/de/diw_01.c.912733.de/nachrichten/energie-_und_klimapolitik_mehr_tempo_ fuer_die_transformation.html.

German Society for Quality. 2024. Digitalisierung in Qualitätsmanagement und Qualitätssicherung. Retrieved November 30, 2024, from: https://www.dgq.de/fachbeitraege/digitalisierung-in-qualitaetsmanagement-und-qualitaetssicherung-beispiele-und-herausforderungen/.

Gilbert, R., & Perl, A. 2018. Transport Revolutions: Moving People and Freight Without Oil. New Society Publishers.

GIZ (Gesellschaft für Internationale Zusammenarbeit). 2017. Sustainable Transport: A Sourcebook for Policy-Makers in Developing Cities. GIZ.

Gomes, S., et al. 2023. Strategic organizational sustainability in the age of Sustainable Development Goals. Sustainability. Retrieved August 04, 2024, from: https://www.mdpi.com/2071-1050/15/13/10053.

Gomez-Baggethun, E., & Naredo, J. M. 2023. Ecological Economics: A Manifesto. Routledge.

Gonzalez-Feliu, J., et al. 2018. Sustainable Urban Logistics: Concepts, Methods, and Information Systems. Springer Publishing.

Göpfert, I. 2024. Logistik - Dynamisch und Innovativ: Führungskonzeption und Management von Supply Chains. Vahlen Publishing.

Grant, D. B. 2023. Environmental Sustainability in Logistics and Supply Chain Management. Kogan Page.

Grant, D. B. 2024. Green Supply Chain Management. Kogan Page.

Grant, D. B., et al. 2017. The handbook of sustainable logistics. Kogan Page.

Greenpeace. 2024. Verkehrswende, für Klima und Gerechtigkeit. Retrieved June 03, 2024, from: https://www.greenpeace.de/klimaschutz/mobilitaet.

Grubb, M., & Neuhoff, K. 2023. Economic pathways to carbon neutrality: Policy and market dynamics. Journal of Environmental Economics and Management, 114, 102469.

Güterbahnen. 2023. Übersicht Zahl des Tages. Retrieved December 09, 2024, from: https://die-gueterbahnen.com/assets/files/2023_12_13-zahl-des-tages.pdf.

H

H2-News. 2024. Wasserstoff per Schiene. Retrieved July 05, 2024, from: https://h2-news.de/wirtschaft-unternehmen/wasserstoff-per-schiene-db-cargo-legt-konzept-vor/.

Experte 5, D. 2024. Expert 5.

Hall, J. M. 2023. Advances in Green Logistics: Emerging Research and Opportunities. IGI Global.

Hall, R. W. 2022. Handbook of Transportation Science (3rd ed.). Springer.

Experte 8, R. 2024. Expert 8.

Harris, N. P., & Williams, J. 2021. Freight transportation, policy measures and market responses. Transport Policy. Retrieved Juli 12, 2024, from: https://www.sciencedirect.com/science/article/abs/pii/S0967070X20309471?via%3Dihub.

Harttmann, C. 2022. Brennstoffzellen LKW kann bis 2030 wettbewerbsfähig sein, Retrieved June 28, 2024, from: https://unterwegs-auf-der-autobahn.de/news/studie-brennstoffzellen-lkw-kann-bis-2030-wettbewerbsfaehig-sein-21237.html

Haug, C. 2023. 2,2 Grad wärmer, Europa erwärmt sich doppelt so schnell wie andere Kontinente. Retrieved June 28, 2024, from: https://www.mdr.de/wissen/copernicus-klimawandel-europa-erwaermt-sich-doppelt-so-schnell-100.html#Durchschnittstemperatur.

Henderson, R. M., & Clark, K. B. 1990. Architectural innovation: The reconfiguration of existing product technologies and the failure of established firms. Administrative Science Quarterly, 35(1), 9-30.

Hensher, D. A., & Button, K. J. 2023. Handbook of Logistics and Supply Chain Management (2nd ed.). Emerald Publishing.

Hensher, D. A., & Mulley, C. 2022. Public Transport: Planning, Management and Operations (3rd ed.). Routledge.

Hettich, G., et al. 2023. Liefer- und Versandverhalten von kommerziellem Frachtgut Nachfrage nach Logistikleistungen Anbieter. In U. Clausen & M. Dellbrügge (Publisher). Interdisziplinären Konferenz für Produktion, Logistik und Verkehr (S. 189-203). Springer Publishing.

Experte 9, K. 2024, Experte 9.

Hinrichs, B. 2021. Nachhaltigkeit als Unternehmensstrategie: Roadmap für unternehmerische Nachhaltigkeit & Innovation. Haufe Publishing.

Hochschule Fulda. 2014. Kausalmodell, Die Suche nach dem wahren Grund. Retrieved June 28, 2024, from: https://www2.hs-fulda.de/~grams/BildungWissenschaft/DerWahreGrund.pdf.

Hoffmann, M. 2021. Nachhaltige Mobilität und Klimaschutz in deutschen Städten. Technische Universität München. Retrieved June 28, 2024, from: https://mediatum.ub.tum.de/doc/1622132/document.pdf.

Hofmann, P. 2023. Grüne Logistik: Ansätze zur Verbesserung der Umweltbilanz im Güterverkehr. Erich Schmidt Publishing.

Hupac Intermodal. 2021. Berichte eines führenden intermodalen Transportunternehmens über die Marktbedingungen und zukünftige Entwicklungen des kombinierten Verkehrs. Retrieved Juli 19, 2024, from: https://www.hupac.com.

Hydrogen Europe. 2025. Hydrogen refuelling projects allocated in latest EU AFIF funding round. Retrieved June 30, 2025, from: https://hydrogeneurope.eu/hydrogen-refuelling-projects-allocated-e33m-in-latest-eu-afif-funding-round/.

HyperloopTT. (2025). First HyperPort Feasibility Study Reveals Transformative Results in Brazil. Retrieved August 28, 2024, from: https://www.hyperlooptt.com/2025/first-hyperport-feasibility-study-reveals-transformative-results-in-brazil/.

hySolutions. 2024. Wasserstoff auf der Schiene. Experte 7 Dr. Simon Experte 7. Retrieved June 16, 2024, from: https://www.hysolutions.de/schiene-wasser-luft/wasserstoff-auf-der-schiene/.

I

IEA. 2025. Renewables. International Energy Agency. Retrieved May 27, 2025, from: https://www.iea.org/energy-system/renewables.

Inal, O. B. 2024. Decarbonization of shipping: Hydrogen and fuel cells legislation in the maritime industry. *Brodogradnja*. Retrieved June 12, 2024, from: https://doi.org/10.21278/brod75205.

Intermodalnews. 2021. Aktuelle Nachrichten und Analysen zur Leistung und Entwicklung des intermodalen Verkehrs in Europa. Retrieved June 13, 2024, from: https://intermodalnews.eu.

International Energy Agency. 2025. Global Energy Review 2025. Retrieved April 28, 2025, from: https://www.iea.org/reports/global-energy-review-2025.

International Transport Forum. 2025. Sustainable Transport Systems Initiative. Retrieved March 28, 2025, from: https://www.itf-oecd.org/sustainable-transport-systems-initiative.

Iskan, S. 2021. Zukunftsbranche Logistik: Zwischen digitaler Industrialisierung und analoger Herausforderung. Campus Publishing.

Izotov, O. 2024. Multi–criteria assessment of the combined shipments container transportation possibility. Retrieved August 09, 2024, from: https://doi.org/10.21821/2309-5180-2024-16-3-413-420.

J

Jackson, E. 2023a. Sustainable Freight Transport and Logistics. Routledge.

Jackson, T. 2023b. Post Growth: Life After Capitalism. Polity Press.

Jakob, M. 2023. Wege zu einer sozial gerechten Verkehrswende: Wie wir Mobilität nachhaltig gestalten können. Hanser Publishing.

Janzen, M., & Erdmann, E. 2024. Wo es in Europa besonders stark gebrannt hat. Retrieved Juli 27, 2024, from: https://www.zeit.de/wissen/2024-06/klimawandel-waldbraende-gefahren-griechenland-studie.

Jöhrens, J., et al. 2022. Vergleichende Analyse der Potentiale von Antriebstechnologien für LKW im Zeithorizont 2030. Institut für Energie und Umweltforschung.

Jones, J., Genovese, A., & Tob-Ogu, A. 2020. Hydrogen vehicles in urban logistics: A total cost of ownership analysis and some policy implications. Renewable and Sustainable Energy Reviews, 119, 109561.

JUSDA. 2024. Emerging Trends in Land Transport for 2025. Retrieved June 20, 2024, from: https://www.jusdaglobal.com/en/article/emerging-trends-land-transport-2025/.

K

Kampp, M. 2022. Herausforderungen im internationalen Schienenfernverkehr in Europa. In Mobilitäts- und Transportrecht in Europa (S. 105-121). Springer Publishing.

Kasten, P. 2022. Klimaschutz im Verkehrssektor. Wirtschaftsdienst. Retrieved June 29, 2024, from: https://link.springer.com/article/10.1007/s10273-022-3169-5

Kaszuba, A., Przybyłowski, A., & Kuzia, M. 2024. Sustainable mobility challenges – Case study of the offshore center in Gdansk transport accessibility. *Economics and Environment*. Available at: https://ekonomiaisrodowisko.pl/journal/article/view/843.

Keller, N. 2020. Klimawandel und Wasserwirtschaft in Deutschland. Universität Karlsruhe. Retrieved Juli 08, 2024, from: https://publikationen.bibliothek.kit.edu/1000118445.

Kerry Health and Nutrition Institute. (2025). Ten key health and nutrition trends of this year. Retrieved March 25, 2025, from: https://khni.kerry.com/trends-and-insights/ten-key-health-and-nutrition-trends-of-this-year/.

Kille, C. 2024. Güterverkehr kompakt: Güterverkehr und seine Bedeutung für Politik, Unternehmen und Märkte (2. Aufl.). De Gruyter Oldenbourg.

Kluth, T., et al. 2022. Gleitende Mittelfristprognose für den Güterverkehr. Retrieved Juli 15, 2024, from: https://www.bmv.de/SharedDocs/DE/Anlage/G/BVWP/verkehrsprognose-2040-zusatzdokument-6-2-prognosepraemissen-prognosefall-1-basisprognose-2040.html.

KOFA Study. 2023. Ökologische Nachhaltigkeit: Mit welchen Kompetenzbedarfen rechnen die Unternehmen?. Retrieved September 14, 2024, from: https://www.iwkoeln.de/studien/paula-risius-susanne-seyda-finn-arnd-wendland-roschan-pourkhataei-monsef-mit-welchen-kompetenzbedarfen-rechnen-die-unternehmen.html.

Koneberg, F., Jansen, A., & Kutz, V. 2022. Energie aus Wind und Sonne: Optionen für die Fachkräftesicherung. KOFA-Studie Nr. 3/2022. KOFA-Kompetenzzentrum Fachkräftesicherung.

König, F. 2021. Klimaschutz in deutschen Mittelstädten (Unveröffentlichte Masterarbeit). Friedrich-Alexander-Universität Erlangen-Nürnberg.

Experte 10, C. 2024. Expert 10.

Krauth, M. 2025. Optimising mode choice in a bi-modal freight network considering sustainability and urban logistic stakeholder perspective. Journal of Transport Geography. Retrieved May 25, 2025, from: https://www.sciencedirect.com/science/article/pii/S2590198225001216.

Krüger, E. 2017. Kommunale Klimaschutzkonzepte in Deutschland. Universität Frankfurt. Retrieved October 01, 2024, from: https://publikationen.ub.uni-frankfurt.de/frontdoor/index/index/docId/44965.

Kummer, S. 2023. Low Carbon Logistics: Reducing Freight Emissions in Transport Chains. Springer Publishing.

L

Laimer, S., & Perathoner, C. 2024. Mobilitäts- und Transportrecht in Europa: Bestandsaufnehme und Perspektiven. Springer Publishing.

Lang, G. T., & Sanchez, M. R. 2021. Innovative technologies for sustainable freight transport. Journal of Advanced Transportation. Retrieved November 17, 2024, from: https://www.researchgate.net/publication/333232430_Innovative_Technologies_for_Sustainable_Passenger_Transport.

Experte 3, G. 2024. Expert 3.

Letta, E., Harper, R., & Zhang, Y. 2024. Enrico Letta on the future of a single European market. *Georgetown Journal of International Affairs*. Retrieved January 13, 2025, from: https://doi.org/10.1353/gia.2024.a934882.

Lexikon der Nachhaltigkeit. 2024. Nachhaltigkeit 2024. Retrieved June 17, 2024, from: https://www.nachhaltigkeit.info/artikel/modelle_und_konzepte_zur_nachhaltigkeit_2018.htm.

Liu, C., et al. 2021. A bi-level model for green freight transportation pricing strategy. Sustainability, 13(63394). Retrieved August 10, 2024, from: https://doi.org/10.3390/su13063394.

Lübben, N., & Kindt, W.-D. 2022. Jahresbilanz 2021 - Zur Lage der deutschen Luftverkehrswirtschaft. Retrieved September 09, 2024, from: https://branchenreport.bdl.aero/application/files/4616/5943/6501/BDL-Jahreszahlen2021_Lagebericht.pdf.

M

Macharis, C., & Melo, S. 2021. Sustainable Logistics and Transportation: Planning and Optimization. Emerald Publishing.

Mainzer, K. 2021. Künstliche Intelligenz und Nachhaltigkeit. In Lecture Notes Informatics (LNI), Gesellschaft für Informatik (S. 85-101). Informatik Publishing.

Experte 4, B. 2024. Expert 4.

Manderscheid, K. 2024. Nachhaltige Mobilität und Wettbewerb: Handbuch zur Umweltsoziologie. Springer Publishing.

Manheller, M. 2024. VDV Jahresbericht. Retrieved January 02, 2025, from: https://www.vdv.de/vdv-jahresbericht-2023-2024.pdfx.

Mariana, S., Luís, S., et al. 2022. Economic, environmental, and social dimensions of freight transportation. Journal of Cleaner Production. Retrieved September 09, 2024, from: https://doi.org/10.1016/j.jclepro.2021.128585.

Experte 6, G. 2024. Expert 6.

Mayer, L. 2019. Klimaschutzmaßnahmen in der deutschen Industrie. Universität Köln. Retrieved May 07, 2024, from: https://kups.ub.uni-koeln.de/10442/.

Mayring, P. 2022. Grundlagen und Techniken der qualitativen Inhaltsanalyse (13. Aufl.). Beltz Publishing.

McKinnon, A. 2018. Transport Logistics: The Future of Sustainable Freight Transport. Kogan Page.

McKinnon, A. 2023. The Future of Sustainable Transport. Edward Elgar Publishing.

Mehmedi, I. 2021. Electric vehicles for urban logistics. Journal of Applied Sciences, 7(13-14), 63-75.

Meißner, S. 2024. Ökologische Logistik: Nachhaltigkeit im Güterverkehr. Springer Gabler.

Merz, H. 2023. Hidden Champions in der Logistik. Dissertation, Nomos Publishing.

Misoch, S. 2019. Qualitative Interviews (2. Aufl.). De Gruyter Oldenbourg.

Experte 12, M. 2024. Expert 12.

Monios, J. 2023. Intermodal Freight Transport and Logistics (2nd ed.). CRC Press.

Monsef, R. P., & Wendland, F. A. 2022. Beschäftigte im Bereich erneuerbare Energien: Produktions- und Fertigungsberufe im Fokus der Energiewende. IW-Report, Nr. 57, Köln.

Muchna, C., et al. 2021. Grundlagen der Logistik: Begriffe, Strukturen und Prozesse (2. Aufl.). Springer Gabler.

Müller, K., Elkins, T., et al. 2024. Green hydrogen: Powering the future of passenger and freight transport. Joule. Retrieved Juli 13 2024, from: https://www.cell.com/joule/fulltext/S2542-4351(21)00453-0?_returnURL=https%3A%2F%2Flinkinghub.elsevier.com%2Fretrieve%2Fpii%2FS2542435121004530%3Fshowall%3Dtrue.

Müller, S. 2020. Klimaschutz und Energiewende in Deutschland. Universität Hamburg. Retrieved Juli 24, 2024, from: https://ediss.sub.uni-hamburg.de/volltexte/2020/10472/.

N

Nagabandi, N. 2023. How 3D printing can dramatically reduce carbon emissions in the manufacturing industry. Manufacturig Tomorrow. Retrieved Juli 19, 2024, from: https://www.manufacturingtomorrow.com/story/2023/05/how-3d-printing-candramatically-reduce-carbon-emissions-in-the-manufacturing-industry/20720/.

Nature Sustainability. 2024. Progress towards the Sustainable Development Goals. Retrieved December 28, 2024, from: https://www.nature.com/collections/bhfffjiadc.

Navoco. 2024. Nachhaltigkeit. Retrieved November 23, 2024, from: https://navoco.de/dreieck-der-nachhaltigkeit-ziele-und-umsetzungen/.

Nell, R. 2024. Mobilität und Logistik, Fraunhofer Forschung. Retrieved June 08, 2024, from: https://www.morgenstadt.de/de/innovationsfelder/MobilitaetLogistik.html.

Nickel, A. 2023. Die Rheinpfalz, Mobilitätskonzept: Jetzt beginnt die Feinarbeit, Retrieved May 07, 2024, from: https://www.rheinpfalz.de/lokal/neustadt_artikel,-mobilit%C3% A4tskonzept-jetzt-beginnt-die-feinarbeit-_arid,5565738.html.

Novocargo. 2024. Last Mile in Urban Areas: The challenge in 2024. Retrieved December 07, 2024, from: https://www.novocargo.com/en/last-mile-in-urban-areas/.

Now. 2024. Elektra II, Wasserstoff Binnenschiff. Retrieved Juli 19, 2024, from: https://www.now-gmbh.de/projektfinder/elektra-ii/.

O

OECD. 2025. The future of global value chains. Organisation for Economic Co-operation and Development. Retrieved Juli 19, 2024, from: https://www.oecd.org/content/dam/oecd/en/publications/reports/2017/07/the-future-of-global-value-chains_9aea7b63/d8da8760-en.pdf.

Öko-Institut. 2024. Faktencheck zur Kernenergie. Retrieved Juli 14, 2024, from: https://www.oeko.de/blog/faktencheck-zur-kernenergie-hochradioaktive-abfaelle-co2-Experte 10en-sicherheit/.

Ordonez-Ponce, E. 2023. Exploring the impact of the Sustainable Development Goals on sustainability trends. Sustainability, 15(24), 16647. https://doi.org/10.3390/su152416647.

Osztovics, W., & Kovar, A. 2021. Arena-Analyse 2021: Marktwirtschaft neu starten. Kovar & Partner Publishing.

P

Page, M. J., et al. 2021. PRISMA source analysis. BMJ, 372, n71. Retrieved August 01, 2024, from: https://www.bmj.com/content/372/bmj.n71.

Palmetto. 2024. The difference between green, clean, and renewable energy. Retrieved November 23, 2024, from: https://palmetto.com/solar/difference-between-green-clean-and-renewable-energy.

Peng, Z. 2024. Trends and Technological Challenges in Green Energy. Springer Publishing.

Perevoznic, F., & Drogomir, V. 2024. Achieving the 2030 agenda: Mapping the landscape of corporate sustainability goals and policies in the European Union. Retrieved August 28, 2024, from: Sustainability. https://doi.org/10.3390/su16072971.

Perez, M. A., & Rossi, I. 2022. Renewable energy integration in freight. Journal of Renewable and Sustainable Energy. Retrieved December 21, 2024, from:

https://www.researchgate.net/publication/382116674_Renewable_energy_integration_in_sustainable_transportation_and_logistics.

Perez, M. A., & Rossi, I. 2022. Renewable energy integration in sustainable transportation and logistics. ResearchGate. Retrieved November 11, 2024, from: https://www.researchgate.net/publication/382116674_Renewable_energy_integration_in_sustainable_transportation_and_logistics.

Pfaff, D. 2020. Praxishandbuch Marketing: Grundlagen und Instrumente (2. Aufl.). Campus Publishing.

Piontek, J. 2021. Bausteine des Logistikmanagements: Supply Chain Management, E-Logistics, Logistikcontrolling, Green Logistics, Logistik Instrumente. NWB Publishing.

Plain Concepts. 2025. Causal ML: Making smarter decisions with causal machine learning. Retrieved May 13, 2025, from https://www.plainconcepts.com/causal-ml/

Q

Quaschning, V. 2018. Regenerative Energiesysteme: Technologie - Berechnung – Simulation (10. Aufl.). Carl Hanser Publishing.

Quentin, P., Buscher, J., & Eltner, T. 2023. Transport planning beyond infrastructural change: An empirical analysis of transport planning practices Germany. Sustainability. Retrieved May 22, 2024, from: https://doi.org/10.3390/su151310025.

R

Racca, R., Ainslie, M. A., Bosschers, J., Hermans, M., Lloyd, T., MacGillivray, A. O., Pace, F., Schuster, M., Sertlek, Ö., & Wood, M. A. 2024. Mapping past and future shipping noise in European seas. *The Journal of the Acoustical Society of America*. Retrieved December 01, 2024, from: https://doi.org/10.1121/10.0027298.

RailFreight.com. 2024. One billion euros for smaller rail infrastructure in Germany by 2025. Retrieved January 18, 2025, from: https://www.railfreight.com/infrastructure/2024/04/05/one-billion-euros-for-smaller-rail-infrastructure-in-germany-by-2025/.

Rail-Pro Alliance. 2023. Gemeinwohlorientierte Infrastruktur: Bund braucht klare Ziele. Retrieved August 12, 2024, from https://www.allianz-pro-schiene.de/presse/pressemitteilungen/gemeinwohl-infrastruktur/.

Rail-Pro Alliance. 2024. Grenzüberschreitender Güterverkehr. Retrieved August 13, 2024, from https://www.allianz-pro-schiene.de/themen/gueterverkehr/.

Railway Technology. 2020. Austria's ÖBB launches hydrogen train passenger trial. Retrieved December 01, 2024, from: https://www.railway-technology.com/news/austria-obb-hydrogen-train-passenger-trial/.

Ramaekers, K., et al. 2018. Using an integrated order picking-vehicle routing problem to study the impact of delivery time windows in e-commerce. European Transport Research Review. Retrieved September 09, 2024, from: https://doi.org/10.1186/s12544-018-0333-5.

Rammert, A. 2024. Mobilität statt Verkehr: Theoretische Grundlagen. Springer Publishing.

Rat der Europäischen Union. 2024. Strategien zur Erreichung der Klimaneutralität bis 2050. Brüssel, EU-Ratsdokumente. Retrieved December 01, 2024, from: https://www.consilium.europa.eu/en/press/press-releases/2024/06/17/soil-monitoring-law-eu-on-the-pathway-to-healthy-soils-by-2050/.

Raza, S. A. 2025. The Potential of Hyperloop in Future Supply Chains. LinkedIn. Retrieved February, 2025, from: https://www.linkedin.com/pulse/potential-hyperloop-future-supply-chains-syed-anwar-raza-bznpc.

Reissova, A., Nachazelova, E., & Cihlar, D. 2024. Job attractiveness and distance as predictors of willingness to move. *TEM Journal*. Retrieved December 01, 2024, from: https://doi.org/10.18421/tem133-36.

REN21. 2024. Global Futures Report: Renewables for Sustainable Transport. Retrieved October 13, 2024, from: https://www.ren21.net/2024-renewables-global-futures-report/.

Research and Markets. 2024. Autonomous Last Mile Delivery Business Analysis Report 2024: Market to Reach $6.2 Billion by 2030. Retrieved May 12, 2025, from: https://www.globenewswire.com/news-release/2025/01/27/3015775/28124/en/Autonomous-Last-Mile-Delivery-Business-Analysis-Report-2024-Market-to-Reach-6-2-Billion-by-2030-Driven-by-Increasing-Use-of-Drones-and-Ground-Robots-Increasing-Adoption-in-Healthca.html.

Reset digital. 2024. Erneuerbare Energien – klimafreundlich und zukunftsfähig. Retrieved December 01, 2024, from: https://reset.org/erneuerbare-energien-umweltfreundliche-und-preiswerte-energie-aus-unerschoepflichen-quellen/.

Reuters. 2025. South Korea's deadly fires made twice as likely by climate change, researchers say. Retrieved April 14, 2025, from: https://www.reuters.com/sustainability/cop/south-koreas-deadly-fires-made-twice-likely-by-climate-change-researchers-say-2025-04-30/.

Richter, F. 2020. Klimaschutz in der deutschen Bauwirtschaft. Universität Stuttgart. Retrieved May 01, 2024, from: https://elib.uni-stuttgart.de/handle/11682/11668.

Experte 13, M. 2024, Experte 13.

Risius, P. 2023. Ökologische Nachhaltigkeit: Mit welchen Kompetenzbedarfen rechnen die Unternehmen? Research Report 2/2023, German Economic Institute.

Rodrigue, J.-P. 2021. The Geography of Transport Systems (5th ed.). Routledge.

Rodrigue, J.-P., Notteboom, T., & Pallis, A. 2022. The Handbook of Logistics and Supply-Chain Management (2nd ed.). Routledge.

Ryley, T. 2023. *Transport and Climate Change: Sustainable Transport in the Making?* Emerald Publishing.

S

Saake, S., & Sommer, C. 2023. Entwurf einer Prognosemethode für die Auslastung im öffentlichen Personennahverkehr auf der Grundlage von Daten aus automatischen Fahrgastzählsystemen. In U. Clausen & M. Dellbrügge (Publisher). Interdisziplinären Konferenz für Produktion, Logistik und Verkehr (S. 233-246). Springer Publishing.

Sampson, H. 2023. Greening the Supply Chain: A Guide for Logistics and Transportation. Palgrave Macmillan.

Savy, M., & Burnham, J. 2023. Freight Transport and the Modern Economy (2nd ed.). Routledge.

Schiffer, H.-W. 2023. Einführung in die Energiewirtschaft: Ressourcen und Märkte. Springer Vieweg Publishing.

Schmidt, P. 2019. Energieeffizienz als Schlüssel zum Klimaschutz. RWTH Aachen. Retrieved November 04, 2024, from: https://publications.rwth-aachen.de/record/776192.

Schmitz, M. 2023. Dekarbonisierung des Güterverkehrs: Wege zu einer klimafreundlichen Logistik. Springer Vieweg.

Schneider, U. 2024. Grüne Transportketten: Effizienzsteigerung und Emissionsreduktion. De Gruyter.

Schultze, J. 2023. Nachhaltige Transportlogistik: Konzepte und Strategien. Springer Gabler.

Seeburger. 2024. Green Logistics – Wie Digitalisierung Nachhaltigkeit vorantreibt. Retrieved December 28, 2024, from: https://blog.seeburger.com/de/green-logistics-wie-digitalisierung-nachhaltigkeit-in-der-logistik-vorantreibt/.

Siebenpfeiffer, W. 2021. Mobilität der Zukunft: Intermodales Verkehrskonzept. Springer Vieweg.

Steininger, K., et al. 2024. Nationaler Energie- und Klimaplan (NEKP) für Österreich - Wissenschaftliche Bewertung der in der Konsultation 2023 vorgeschlagenen Maßnahmen. CCCA.

Stelling, P., & Brunner, S. 2023. Erforschung von Wissenslücken zwischen den Hauptakteuren beim Übergang zu einem elektrifizierten Güterverkehrssystem in Schweden. In U. Clausen & M. Dellbrügge (Publisher). Interdisziplinären Konferenz für Produktion, Logistik und Verkehr (S. 259-272). Springer Publishing.

Stevens, V. E., et al. 2023. Advancements in electric and hydrogen fuel cell vehicles for freight transport. Renewable and Sustainable Energy Reviews. Retrieved June 04, 2024, from: https://doi.org/10.1016/j.rser.2022.112499.

Stölzle, W., Schöder, D., & Häberle, L. 2024. Güterverkehr kompakt (2. Aufl.). De Gruyter Oldenbourg.

Experte 1, M. 2024. Expert 1.

System Bahn. 2024. Schienenfernverkehr in Europa, Status Quo, Herausforderungen und Perspektiven. Retrieved June 17, 2024, from: https://www.system-bahn.net/aktuell/schienenfernverkehr-in-europa-status-quo-herausforderungen-und-perspektiven/.

Szaruga, E., Załoga, E., Drewnowski, A., Kowalska, S., & Dąbrosz-Drewnowska, P. 2024. The role of EU transport market liberalization in shaping directions of rail energy consumption rationalization in relation to the export of goods: The case of Poland. *Energies, 17*(13), 3118.

T

Tagesschau. 2024a. Wasserstoff für die Schiene. Redakteur: Thomas Denzel. Retrieved October 16, 2024, from: https://www.tagesschau.de/wirtschaft/unternehmen/wasserstoff-deutsche-bahn-100.html.

Tagesschau. 2024b. Ein Beitrag von Till Bücher, Finanzdirektion, so läuft der Ausbau der Erneuerbaren Energien. Retrieved November 19, 2024, from: https://www.tagesschau.de/wirtschaft/energie/erneuerbare-energie-ausbau-stromerzeugung-100.html.

Texas A&M Transportation Institute. 2023. Urban Mobility Report 2023. Retrieved June 22, 2024, from: https://tti.tamu.edu/news/tti-publishes-2023-urban-mobility-report/.

Thakur, T. 2025. Rising network bandwidth: 2025 trends and best practices. Broadband Communities Magazine. Retrieved May 25, 2025, from: https://bbcmag.com/rising-network-bandwidth-2025-trends-and-best-practices/.

The Sustainable Agency. 2025. Greenwashing examples for 2024 & 2025. Retrieved May 24, 2025, from: https://thesustainableagency.com/blog/greenwashing-examples/.

Tovias, A. 2024. Will the EU's fourth freedom be further challenged by present members of the single market? *Central European Economic Journal, 11*(42), 42–53. Retrieved December 11, 2024, from: https://doi.org/10.2478/ceej-2024-0005.

Tumlin, J. 2024. Sustainable Transportation Planning: Tools for Creating Vibrant, Healthy and Resilient Communities. Wiley Publishing.

U

UBA Dessau. 2023. Die UBA-Prognose, THG sanken um 1,9 Prozent. Umweltbundesamt. Retrieved June 30, 2024, from: https://www.umweltbundesamt.de/presse/pressemitteilungen/uba-prognose-treibhausgasemissionen-sanken-2022-um.

UIC Railway Publications. 2021. Berichte zur Rolle des kombinierten Verkehrs im Rahmen des Europäischen Green Deals und zur Reduzierung von CO2-Emissionen. Retrieved May 19, 2024, from: https://uic.org/IMG/pdf/uic_activity_report_2020-2021.pdf.

UIRR - International Union for Road-Rail Combined Transport. 2020. Der Bericht 2020 über den kombinierten Verkehr in Europa. Retrieved May 19, 2024, from: https://uic.org/IMG/pdf/2020_report_on_combined_transport_in_europe.pdf.

Umbach, F. 2023. Energiesicherheit unter Bedingungen der Dekarbonisierung von Wirtschaft und Verkehr. Sirius. Retrieved May 22, 2024, from: https://www.degruyter-brill.com/document/doi/10.1515/sirius-2023-2002/html.

Umweltbundesamt. 2022. Digitale Nachhaltigkeit. Retrieved May 19, 2024, from: https://www.umweltbundesamt.de/themen/digitalisierung/digitale-nachhaltigkeit.

Umweltbundesamt. 2022. Minderung der THG, Ziele Deutschlands. Retrieved May 18, 2024, from: https://www.umweltbundesamt.de/daten/klima/treibhausgasminderungsziele-deutschlands#nationale-treibhausgasminderungsziele.

Umweltbundesamt. 2023. Forschungsprogramm des Umweltbundesamt 2023-2027. Der europäische Green Deal und seine Auswirkungen auf die deutsche Klimapolitik. Dessau-Roßlau, Deutschland: UBA-Berichte. Retrieved June 06, 2024, from: https://www.umweltbundesamt.de/publikationen/forschungsprogramm-des-umweltbundesamtes-2023-2027.

Umweltbundesamt. 2024. Erneuerbare Energien im Verkehr. Retrieved October 16, 2024, from: https://www.umweltbundesamt.de/daten/energie/erneuerbare-energie-im-verkehr.

Umweltbundesamt. 2024. Klimaschutz im Verkehr. Retrieved May 19, 2024, from: https://www.umweltbundesamt.de/themen/verkehr/klimaschutz-im-verkehr.

Umweltbundesamt. 2024a. Digitale Nachhaltigkeit, Digitalisierung nachhaltig gestalten. Retrieved June 07, 2024, from: https://www.umweltbundesamt.de/themen/digitalisierung/digitale-nachhaltigkeit.

Umweltbundesamt. 2024b. Digitale Mobilitätsplattformen: Analyse und Bestandsaufnahme zu plattformbasierten Mobilitätskonzepten. Umweltbundesamt.

Umweltbundesamt. 2024c. Ziele des Pariser Klima-Abkommens. Retrieved June 11, 2024, from: https://www.umweltbundesamt.de/themen/klimaenergie/internationale-eu-klimapolitik/uebereinkommen-von-paris#ziele-des-ubereinkommens-von-paris-uvp.

Umweltbundesamt. 2024d. Klimaschutz im Verkehr. Retrieved June 12, 2024, from: https://www.umweltbundesamt.de/themen/verkehr/klimaschutz-im-verkehr#undefined.

Umweltbundesamt. 2024e. Umweltmonitor 2024 – Daten zur Umwelt, Publisher Dirk Messner, Dessau.

Umweltbundesamt. 2024f. Lokaler, nationaler und internationaler umweltschonender Güterverkehr – Datenbericht zum Transformationsdialog und UBA Forum mobil und nachhaltig. Texte 89/2024, Dessau.

UN environment. 2021. Green Energy Choices: Vorteile, Risiken und Kompromisse kohlenstoffarmer Technologien für die Stromerzeugung, Retrieved May 28, 2024, from: https://www.resourcepanel.org/reports/green-energy-choices-benefits-risks-and-trade-offs-low-carbon-technologies-electricity.

United Nations Department of Economic and Social Affairs. (2025). World Social Report 2025: A New Policy Consensus to Accelerate Social Progress. Retrieved May 26, 2025, from: https://desapublications.un.org/publications/world-social-report-2025-new-policy-consensus-accelerate-social-progress.

Unsere Welt in Daten. 2023. Treibhausgasemissionen nach Sektoren. Retrieved July 20, 2024, from: https://ourworldindata.org/grapher/ghg-emissions-by-sector?time=latest.

V

Varnavskii, V. G. 2024. EU railways: Contribution to decarbonisation solution. *Contemporary Europe*. Retrieved December 19, 2024, from: https://www.researchgate.net/publication/387070924_EU_Railways_Contribution_to_Decarbonisation_Solution.

VCD. 2022. Güterverkehr gefährdet das Klima. Retrieved May 19, 2024, from: https://www.vcd.org/artikel/gueterverkehr-gefaehrdet-das-klima.

VDE. 2022. VDE /VDI Studie 2022. Retrieved July 14 2024, from:: https://www.vde.com/resource/blob/2208680/856e76b3bf2c31104030c12127e0f0bd/wasserstoff-fuer-den-schienenverkehr-data.pdf.

VDI. 2024.Logistikinitiative will mehr Fracht auf Schiene und Binnenschiff bringen, Verband Deutscher Ingenieure. Retrieved July 14, 2024, from: https://www.ingenieur.de/fachmedien/vdi-z/logistik/transportlogistik/logistikinitiative-will-mehr-fracht-auf-schiene-und-binnenschiff-bringen/.

VDV. 2024. Dem Schienenverkehr Priorität geben. Retrieved July 22, 2024, from: https://www.vdv.de/jetzt-mehr-bewegen-schienengueterverkehr.aspx.

Verkehrsclub. 2023. Bund für Natur und Umweltschutz in Deutschland, Wahnsinn Güterverkehr. Retrieved May 03, 2024, from: https://www.vcd.org/fileadmin/user_upload/Redaktion/Themen/Gueterverkehr/Lkw-Maut/BUND_VCD-Flyer_Gueterverkehr_klein_1.pdf.

Experte 7, S. 2024. Expert 7.

VKU. 2024. Verband kommunaler Unternehmen, mögliche Auswirkungen eines EU-Klimaziels von -90% für 2040 auf Deutschland, Retrieved June 10, 2024, from: https://www.vku.de/ fileadmin/user_upload/Presse/240929_VKU_DIHK_Studie_Klimaziele_2040_deutsch.pdf.

von Malmborg, F. 2024. At the controls: Politics and policy entrepreneurs in EU policy to decarbonize maritime transport. *Review of Policy Research*. Retrieved October 13, 2024, from: https://doi.org/10.1111/ropr.12609.

W

Waalany. 2024. PESTEL Analyse. Retrieved October 13, 2024, from: https://blog.waalaxy.com/de/pestel-analyse/.

Wagner, B. 2023. Nachhaltige Güterverkehrskonzepte: Planung und Umsetzung. Springer Gabler.

Wagner, H., & Beschler, T. 2024. Mobilität im Wandel der Zeit. Springer Publishing.

Weber, I. 2019. Erneuerbare Energien und Netzstabilität in Deutschland. Universität Dresden. Retrieved October 13, 2024, from: https://tud.qucosa.de/api/qucosa%3A36375/attachment/ATT-0/.

Weber, T. 2023. Nachhaltigkeitsmanagement in der Logistik: Praxisbeispiele und Strategien. Erich Schmidt Publishing.

Weidemann, H.-J. 2022. Emissions- und staufreier alpenüberquerender Güterverkehr: Wachstumspotenziale und Hürden für den Verkehrsträger Schiene am Beispiel „Brenner-Pass". In Mobilitäts- und Transportrecht in Europa (S. 299-314). Springer Publishing.

Wensauer, H.-P. 2023. Nachhaltige Logistik: Strategien für eine ökologische Zukunft. Springer Gabler.

Wieland, A., & Durach, C. F. 2021. Two perspectives on supply chain resilience. *Journal of Business Logistics*, 42(3), 315-322. Retrieved June 24, 2024, from: https://doi.org/10.1111/jbl.12271.

Wilde, M. 2023. Neue Mobilität: Faktoren, Treiber und Veränderungen. Springer Publishing.

Wired. 2023. The EU Just Kicked Off Its Biggest Climate Experiment Yet. Retrieved August 20, 2024, from: https://www.wired.com/story/eu-carbon-tax.

Witte, A. 2024. Warum die Schiene dem LKW kaum Konkurrenz macht. Retrieved October 12, 2024, from: https://www.arnewitte.de/warum-die-schiene-dem-lkw-kaum-konkurrenz-macht/.

World Economic Forum. 2025. Clean energy is key to building resilience in uncertain times. Retrieved May 20, 2025, from: https://www.weforum.org/stories/2025/04/clean-energy-resilience-uncertain-times/.

World of Work. 2024. The Henderson-Clark Innovation Model: A Simple Summary. Retrieved January 12, 2025, from: https://worldofwork.io/2019/07/the-henderson-clark-innovation-model/.

Wozniak-Jechorek, B. 2024. Examining the European transition: Central Eastern European countries two decades post-EU accession. *Revista Economia e Políticas Publicationss*. Retrieved December 12, 2024, from: https://doi.org/10.46551/epp2024v12n0104.

Wright, R. E. 2023. Sustainable Freight Transport: Theory, Modelling and Case Studies. Edward Elgar Publishing.

Wu, L., Liu, S., Qi, L., & Lin, D. 2024. Mandatory disclosure and corporate ESG performance: Evidence from China's "explanation for nondisclosure" requirement. *Corporate Social Responsibility and Environmental Management*. Retrieved October 30, 2024, from: https://doi.org/10.1002/csr.2944.

X

Xiong, X., Xing, X., Li, H., Xue, R., Li, Y., & Liu, T. 2024. Calculation of mixed oil volume for sequential transportation of crude oil pipelines with variable temperature and low Reynolds number. *Journal of Physics: Conference Series, 2731*(1). Retrieved January 14, 2025, from: https://doi.org/10.1088/1742-6596/2731/1/012043.

Y

Yunex Traffic. 2025. *Trend Report 2025: ITS Industry in the Intelligent Age*. Retrieved April 20, 2025, from: https://www.yunextraffic.com/newsroom/trend-report-2025-its-industry/.

Z

Zielinski, S. 2024. *Green Transportation: Technologies, Policies, and Practices*. Elsevier Publishing.

Appendix

Appendix 1: AI tools directory

ChatGPT-4, 2025

Perplexity.ai, 2025

Napkin.ai, 2025

Appendix 2: Word Cloud

Figure 18: Word Cloud
Source: Own representation.

A word cloud is a visual representation of textual data, where the frequency or significance of words is emphasized through their size and, occasionally, their color. Representation of the Word Cloud with the keywords relevant to this work.

Appendix 3: Interview guide

Interview guide for the expert interviews on the topic

"Optimizing the sustainability of European freight transport"

- Green Energy in the mobility competition -

Expert interview from: ______________________
Place, date, time

Name of the interviewee: ______________________
Title, first name, last name

In the function: ______________________
Relevance to the work

Before the interview

a. Giving thanks	Gratitude for the willingness to participate
b. Introduce yourself	Name, studies, company, professional function
c. Introduction of an expert	Company, position, function, etc.
d. Description of the dissertation	Research objective, theory, practice, empiricism
e. Procedure	Interview approx. 45 min, free narration, there are no wrong answers, the expert's point of view counts
f. Confidentiality	Data protection Audio recording, anonymization of personal data if desired
g. Anonymization	Is anonymization desired? [ ] yes [ ] no
h. Questions	Does the interviewee have any questions? [ ] yes [ ] no

Starting the recording

A.1 Getting started

- Gender: Experience in years:
- Country: Age:

B. Potentials and risks

- B1. Where do you see the greatest potential for increasing sustainability?
- B2. Where do you see the greatest risks in increasing sustainability?

C. Ecological sustainability

- C1. How can European freight transport ensure environmental sustainability in the future?
- C2. What needs to be improved in the construction of infrastructure and the development of drives?

D. Social sustainability

- D1. Which aspects will play a greater role in European freight transport than they do today?
- D2. What measures and incentives could there be to increase social acceptance?

E. Economic sustainability

- E1. How can freight transport ensure economic sustainability?
- E2. The load capacity of the infrastructure has already been reached. What economic measures could relieve the strain on the infrastructure in the future?

F. Technical sustainability

- F1. How could technology change in the next few years?
- F2. How can IT or AI solutions increase sustainability?

G. Green Energy

- G1. What incentives could politicians create to encourage the use of green energy in the future?
- G2. How do infrastructure, technology and drives need to change in order to produce less greenhouse gases for future generations?

H. Exit

- H1. Are there any other points you would like to talk about? Thank you very much!

Ending the recording

After the interview

i. Time Space	Is the expert ready for another round of discussion ?
j. Thanking	the interviewee again for their willingness to participate and for their time.
k. Questions	Does the interviewee have any final questions?
l. Farewell	Say goodbye and give feedback

Appendix 4: Structure of the work

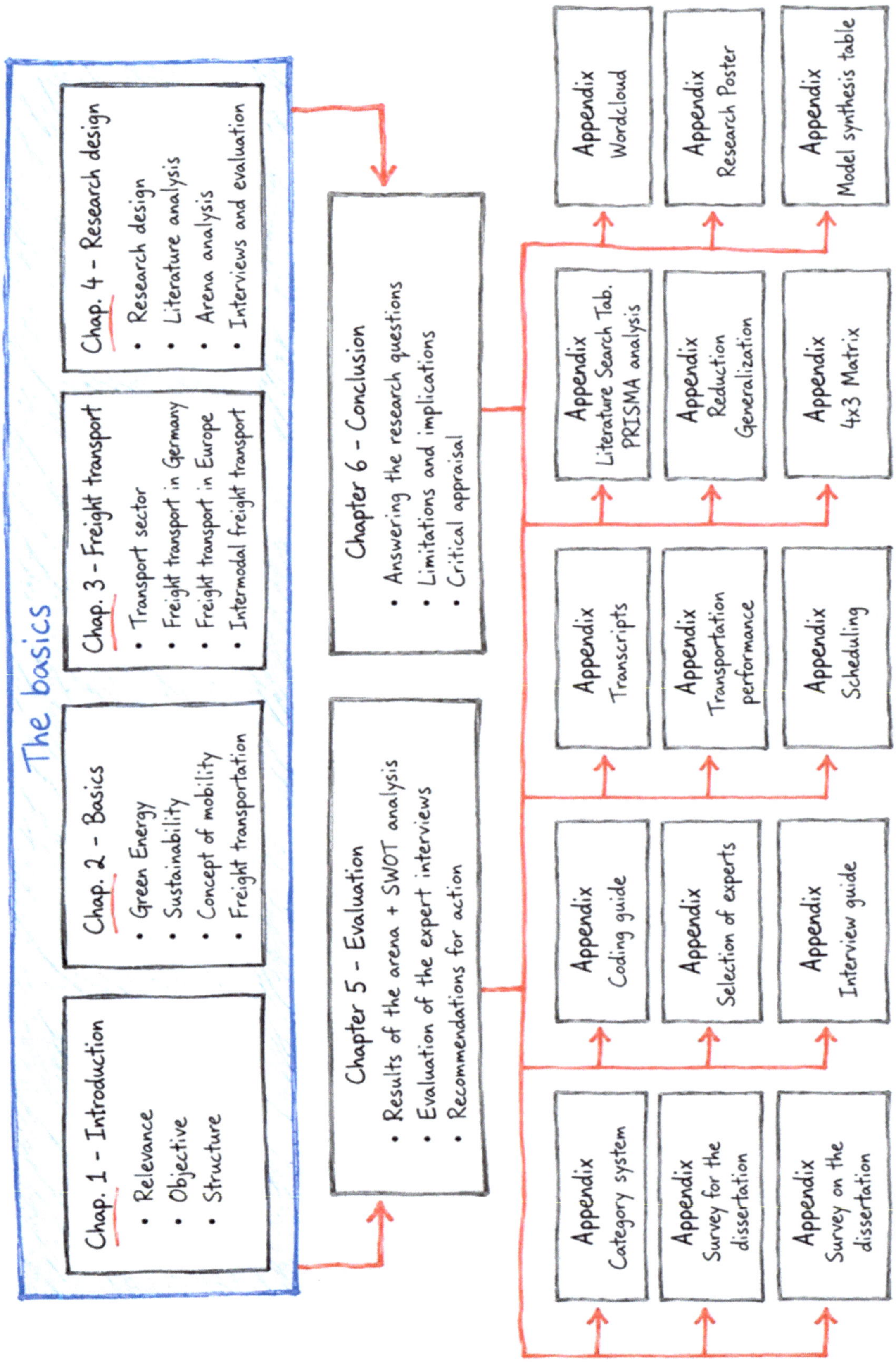

Figure 19: Strukture of the work
Source: Own representation.

Appendix 5: List of the expert interviews conducted

No.	Expert	Date	Duration PLAN	Duration IS
1	Experte 2 (Ob Ovo)	05.07.2024	45 min	58 min
2	Experte 1 (DHL)	02.07.2024	45 min	43 min
3	Experte 3 (VDV)	11.07.2024	45 min	64 min
4	Experte 9 (CGI)	27.08.2024	45 min	29 min
5	Experte 4 (Deloitte)	15.07.2024	45 min	41 min
6	Experte 5 (Pro-Rail Alliance)	16.07.2024	45 min	31 min
7	Experte 6 (Schenker)	23.07.2024	45 min	63 min
8	Experte 7 (hySolutions)	26.07.2024	45 min	49 min
9	Experte 8 (Deutsche Bahn)	01.08.2024	45 min	52 min
10	Experte 10 (Fraunhofer)	13.09.2024	45 min	53 min
11	Experte 11 (Swiss Federal Railways)	10.09.2024	45 min	38 min
12	Experte 12 (Austrian Federal Railways)	12.09.2024	45 min	53 min
13	Markus Experte 13 (Brenntag)	13.09.2024	45 min	32 min
Total: 13 Interviews conducted			**585 min 9,75 h**	**601 min 10,01 h**

Table 17: Proof of the expert interviews conducted
Source: Own representation.

Appendix 6: Deductive category system

Cat.	Deductive categories	Source	Cat.	Inductive Sub-categories
C1	Potentials and risks of mobility	Wagner, H. and Beschler, T. (2024): Mobility through the ages, Springer Publishing.	C1.1 C1.2 C1.3 C1.4 C1.5 C1.6	Shift to rail Mobility Technology Combined transport Electrification Combined transport
C2	Ecological sustainability	Kummer, S. (2023): Low Carbon Logistics, Reducing Freight Emissions in Transport Chains, Springer Publishing, Berlin.	C2.1 C2.2 C2.3 C2.4 C2.5 C2.6 C2.7	Rail is sustainable Mobility Combustion engines Construction of terminals Alternative Cargo Loop Develop. and refurbishment
C3	Social Sustainability	Göpfert, I. (2024): Logistik - dynamisch und innovativ: Führungskonzeption und Management von Supply Chains, Vahlen Publishing Munich.	C3.1 C3.2 C3.3 C3.4 C3.5	Rethinking Traveling personnel Competitiveness Mobility Expansion
C4	Economic sustainability	Grant, D. B. (2023): Environmental Sustainability in Logistics and Supply Chain Management, Kogan Page Publishing, London.	C4.1 C4.2 C4.3 C4.4 C4.5	Company image Financing Importance of logistics Relief for the infrastructure Economic efficiency
C5	Technical sustainability	Mainzer, K. (2021): KI und Nachhaltigkeit, in Lecture Notes Informatics (LNI), Gesellschaft für Informatik, ed. computer science, Bonn.	C5.1 C5.2 C5.3 C5.4 C5.5 C5.6	3D printing Technical changes IT+ Digitization Optimization potential Exchange of information AI solutions
C6	Green Energy	Schneider, U. (2024): Grüne Transportketten, Effizienzsteigerung und Emissionsreduktion, De Gruyter Publishing, Berlin.	C6.1 C6.2 C6.3 C6.4 C6.5	Control system Increase efficiency Raw materials Rethinking Drives and energy sources
C7	More Aspects		C7.1 C7.2	Cooperation in the EU Increase speed

Table 18: Category system
Source: Own representation.

Appendix 7: Coding guide and coding rules

Cat.	Coding	Definition	Anchor example	Coding rules
C1	Potential and risks	Potentials and risks refer to possible benefits and risks associated with a decision or development that influence future results.	"The biggest lever in more efficient transport planning lies in electrification", (expert interview 5 Experte 4, (2024), line 43).	Example: Hydrogen on the road can be a solution for heavy traffic.
C2	Ecological Sustainability	Use resources in such a way that ecological systems retain their ability to regenerate in the long term and future generations can benefit.	"Convert drives so that sustainable means of propulsion can be used", (Expert interview 6 Experte 5, (2024), line 101).	Example: Switch to green fuels.
C3	Social Sustainability	Shaping social structures so that they promote social justice, equal opportunities and cohesion in the long term.	"Automation creates jobs that are pleasant for people." (Expert interview 8 Experte 8, (2024), line 270).	Example: Social standards must rise.
C4	Economic Sustainability	Ensuring an economic system that uses resources efficiently and enables economic growth without jeopardizing social and ecological foundations.	"The truck companies can use CT and thus provide explosive capacities", (Expert interview 3 Experte 3, (2024), line 318).	Example: Improve CO_2 balance.
C5	Technical Sustainability	Developing technological systems in such a way that they conserve resources and maximize the benefits for people and the environment over their entire life cycle.	"Through automation, digitalization and faster communication", (Expert interview 2 Experte 1, 2024), line 142).	Example: Introduction of digital automatic coupling.
C6	Green Energy	Green energy refers to energy that is generated from renewable sources such as sun, wind, water, geothermal heat and biomass and has a minimal negative impact on the environment.	"Convert diesel locomotives to hybrids so that locomotives can run on electricity and hydrogen", (Expert interview 4 Experte 9, (2024), lines 266-267).	Example: Reduce existing subsidies for fossil fuels.
C7	More Aspects	None	none	Example: none

Table 19: Coding guide and coding rules
Source: Own representation.

Appendix 8: Paraphrasing and reduction I of transcripts

Block B Potentials and risks	Interview partner (statements and line number)												
Statement Reduction I	Experte 2	Experte 1	Experte 3	Experte 9	Experte 4	Experte 5	Experte 6	Publisher	Experte 8	Experte 10	Experte 11	Experte 12	Experte 13
The potential lies in the collective load		12											
Shift traffic to rail	16	13	20				28	192		159	118		
Relieving the strain on infra-structure and ca-pacities	24	20											
There used to be many more goods connections		22											
Rail is the most sustainable	18												
Infrastructure is outdated and does not meet the requirements	23												
Freight transport should work more digitally	25			33									
Accelerate the ex-change of infor-mation	26												
In reality, there is no switch to rail			25										
People must be prepared to ac-cept change			36										
to use available natural resources more consciously			45										
We have to buy energy			63										
People should use resources wisely			67										
The biggest lever in more efficient transport plan-ning lies in elec-trification					43			37				24	

Hydrogen is an alternative					44		38						
Companies are interested in investing in digital solutions					52								
Switching to green fuels					61								
e-Fuels could be a solution					63		35						
Many small measures will bring improvements						40							
People should adopt sustainable behavior						46							
Companies have concerns about negative effects						58							
Companies shy away from investments						60							
Companies bear reputational risks						64							
Sustainability is an important topic							22					29	
Sustainability is good, must not cost anything							31						
it will not be possible to consistently shift air freight to rail							44						
emissions from air freight by switching to rail.							48						
infrastructure and availability of large quantities of green hydrogen							66						
Raw materials for batteries must be extracted from the earth							77						

Hydrogen on the road can be a solution for heavy goods traffic								40					
Using hydrogen on the railways where electrification is no option								41					
Hydrogen imports cannot be ramped up quickly enough								52					
Freight transport could be greener									53				
Trains can feed energy back into the grid									59				
Building energy storage									75				
Replacing all diesel locomotives is too cost-intensive									78				
Optimization through digitalization				34					87				
Introduction of DAC is important, currently 70,000 manual clutches per day				200					89			126	
Coupling must be automated									95				
Energy is required for acceleration and braking									105				
Solution must be coordinated at European level									123				
Complete electrification is not possible for economic reasons									132				
Sustainability means rail technology, climate, nature, resources and noise protection									136				
Noise also pollutes the environment									141				

New approaches with quiet wheel-sets									142				
Linking different modes of transport more closely together				34									
Combine transports				37									
The infrastructure was neglected				53									
Equivalent infrastructure Transport modes missing				60									
the provision of energy becomes so expensive that traffic is burdened											31		
if you don't have any money, you will also be pushed to the back politically											46		
Infrastructure is designed for sustainability												36	
Renewable energies in the energy mix												32	
Financing risk issues is difficult to quantify												47	
Sustainability must first be transparent												55	
in energy generation as such													13
that for us, getting bogged down means putting all our eggs in one basket													24
Avoid greenwashing													26
the industry then tends to overdo it or exaggerate it													27

is not even planted especially for this purpose, but also cleared													30
Electricity mix and no certificate purchases										19			
Electrify freight transport										50			
Road-bound truck traffic, due to high sustainability potential										57			
the risk potential and uncertainty										65			

Block C Ecological sustainability	Interview partner (statements and line number)												
Statement - Reduction I	Experte 2	Experte 1	Experte 3	Experte 9	Experte 4	Experte 5	Experte 6	Publisher	Experte 8	Experte 10	Experte 11	Experte 12	Experte 13
Alternat. Use firing carrier		30											
it is difficult to operate an entire truck with solar energy		44											
Hyper-Loop / Cargo-Loop	50												
Underground installation	52												
Build more sustainable infrastructure	69				123	100							
A holistic view of the market is crucial	71												
What to do with batteries that need replacing	80												
Landscapes have to be destroyed again to extract lithium	82												
Should be affordable			91										
the same state as the 60s			100										

There are still many single-tracks, trains have to share a track in each direction			101										
Rural areas not electrified			107										
freight traffic needs its own network			116										
Infrastructure, especially the rail network, needs to be completely renovated			128										
Building more terminals for multimodal terminals			132										
sometimes drive 20 km on battery power if there are no overhead lines			146										
Think about the first and last mile			170										
Think about sustainable and cost-effective options					106								
speak not only of decarbonization, but of ecological behavior					112								
Research and development of drives					123								
e-Fuels contribute to ecological sustainability					130								
Corresponding pipelines are required for hydrogen					131								
100 percent renewable power						81							
drives, such as sustainable drive systems like methanol and hydrogen						101							

Adequate supply of drive units						104							
long distances by rail							90						
Mixture of technologies more interesting							92						
Small distances mean fewer emissions, as there is less wind resistance							95						
the government should create incentives							99						
generate alternative drive systems and then follow up with infrastructure							115						
Overall view							127						
away from fossil fuels								61					
There is a very high level of development activity and potential								74					
Electrify as much as possible								76					
Work and research must be carried out on fuel cells								80					
There are long approval processes for research								85					
Hydrogen import and distribution capacities must be ensured								87					
Companies are hesitant because the business case is still missing								99					
Walk the entire train and check every brake on every freight train									156				

Ensure braking and downhill control									161				
The freight trains are put together using remote-controlled locomotives									165				
There are many track widths in Europe									183				
Transhipment points need to be modernized									192				
Optimizing through automation and digitalization									201				
Huckeplack traffic is simple and straightforward									223				
Development of new inverter technology									240				
Battery technology needs to be improved									243				
Batteries and hydrogen are not yet suitable for high outputs									245				
Reduce the use of fossil fuels				69									
Openness to technology				94									
Consideration of the production chain				95									
Creating public consensus				103									
Merger of alliances											55		
build up the process in such a way that there are no major losses											57		

Creating a competitive advantage											59		
Use the energy correctly											61		
Bring the entire infrastructure up to date											67		
renew the existing lines											69		
Infrastructure of connection and route must be realized												66	
Establish a pan-European network												71	
Create the basis for realizing freight transport												72	
Problems with recrewing the locomotives												76	
Station planning also increases attractiveness												93	
Material selection is a major aspect of sustainability												98	
Do not seal the soil completely												100	
Noise barriers must be designed in such a way that they are beneficial to the environment												102	
They want to move away from diesel locomotives and more towards electric mobility												107	
CO_2 imprint, which can be clearly defined and is definitely not uninteresting													43
We should not limit ourselves													57

Logistics planning, avoid empty runs										85			
the pure purchase of certificates is really not the 100% environmental dimension that really helps us										87			
to rely on ecological technology										90			
What conventional energy sources will develop over the next 20 years										109			
Block D Social sustainability	Interview partner (statements and line number)												
Statement - Reduction I	Experte 2	Experte 1	Experte 3	Experte 9	Experte 4	Experte 5	Experte 6	Publisher	Experte 8	Experte 10	Experte 11	Experte 12	Experte 13
Salaries of drivers and warehouse employees are insufficient		54		126		115							
Train drivers are now well paid		57											
Truck drivers must be able to spend a reasonable night on the road		58											
Create more charging points for e-vehicles		70											
There is far too little loading capacity		75											
Charging time of 2-3 hours is too long		79											
via induction		86											
Routing across the floor		87											
bring in the human aspect and reason	105												
Our society must change	112												

will be more digital on the road	119												
Simple and intuitive	132												
think more about the users	140												
Tractors without driver's cab			198										
Rest areas are not hotels, they are simply parking lots			205										
Building more truck stops			213										
Foreign companies continue to operate below the minimum wage			255										
Massive loads from CT are going onto the road - this should be counteracted			270										
Funds are made available for the expansion of the infrastructure			284										
Funding for the construction of multimodal terminals			287										
Most people are not ready for change			296										
We have both a democratic and a democratic change					151								
Employers must take care to ensure fairness					156								
compliance with the ethical principles					159								
Market regulation through bans					177								
what we also have to learn to					185								

do without as human beings													
Fluctuation is very high						121							
There needs to be more regulation in the social sphere						122							
not competitive with the market in many areas						130							
Social standards must rise				130		133							
either you force someone to do it						145							
or on a voluntary basis with self-commitment						147							
especially state subsidies							143						
The effect of a salary increase is forgotten within three months							144						
Waiving taxes or support is the biggest lever							149						
green and dirty electricity							157						
Buying green transportation solutions							160						
People find it difficult when it comes to doing without							167						
Other incentive systems								117					
Subsidies								119				170	
alternative technologies are competitive in many areas								125					
Anyone who wants to drive 1000 km now has a problem								126					

Automation eliminates jobs that nobody wants to do									263				
Automation creates jobs that are pleasant									268				
Nobody wants to do heavy shunting jobs any more									270				
Noise pollution must be reduced									280				
Trucks and transshipment points produce noise and need to be optimized									284				
Import of labor and thus distortion of competition				124									
Equalizing wage dumping				127									
Creating awareness about hourly wages and working conditions				149									
Payment and social benefits for truck drivers											78		
Action is necessary, more control is needed											79		
Put a stop to injustice											92		
Relieving employees of heavy loads												128	
Hazardous substances that employees come into contact with												131	
Be aware of the dangers of cleaning products												135	
digitize through intelligent systems, imaging procedures												145	

On-site use due to VDU work-station												146	
simply show transparently what effects our actions have												155	
So here is the traceability and it must be shown transparently													76
Gaining accep-tance													79
Avoid green-washing										145			
Infrastructure measures to com-municate envi-ronmental dimen-sions										163			

Block E Economic sustainability	Interview partner (statements and line number)												
Statement - Re-duction I	Experte 2	Experte 1	Experte 3	Experte 9	Experte 4	Experte 5	Experte 6	Publisher	Experte 8	Experte 10	Experte 11	Experte 12	Experte 13
Tax concessions are one solution	158	96											
The load capacity and limits of the infrastructure have been reached	179	103											
Increase road toll		105											
use a portion of the fuel tax		106											
depending on consumption		108											
also introduce road tolls for cars		109											
Create incentives	155												
The infrastructure is there, and there are also enough routes	162												
There must be enough wagons available to get them out of the depot quickly enough	180												

It is also not clear whether the goods will actually arrive on time	182												
Truck companies can use CT and thus provide explosive capacities			318										
The optimal solution is a clever mix of truck, rail and other modes of transportation.			330	34									
Keeping an eye on how track access charges are developing			367										
make sure that cup prices do not go up			374										
On some routes there is one train per hour and then we are already talking about capacity bottlenecks			393										
Promoting digital interlocking systems			410										
provide a continuous freight transport route			417										
Logistics is the backbone of a modern society					203								
Logistics per se ensures prosperity					209								
Logistics must remain affordable					212								
not on the backs of the blue-color workers (working class)					214								
with a high taxation rate					228								
State or political task					235								
continue to try to cover the costs of their suppliers						161							

and increase prices to their customers													
they can try to lower suppliers' prices through market power						165							
Freight transport offers a lot of stability and spontaneity							186						
transport chains of high-value goods or are always on the move for long distances							191						
You shouldn't necessarily reduce the number of trucks and drivers, but rather the space that a truck needs.							208						
Alternative: Larger trucks - Gigaliner - Hub transports							209						
small electric vehicles instead of the large panel vans							218						
Freight transport is the backbone of the economy								137					
enable production processes								139					
A shift in the direction of rail is certainly the								157					
The overall burden on infrastructure should be reduced								165					
to bring more traffic onto the railways								166					
Advancing the expansion of rail								168					
Automation and control systems									440				

make the infrastructure more efficient													
Have everyone drive behind each other at an optimized speed									444				
Allowing vehicles with the same power to drive in one lane optimizes the flow									455				
Create basic infrastructure				172									
The state must enable competition in infrastructure				177									
Optimization through route utilization				186									
Regulation via variable fees				187									
Trying to manage demand				193									
map long traffic flows											107		
Form more alliances to offer more transport connections											110		
the truck is not allowed to drive on all roads in Switzerland											119		
Relief on the roads											121		
Avoid permanent construction site											131		
Impose a CO_2 tax on freight transport and make prices more transparent												188	
The last mile will remain difficult												192	
offer competitive and attractive												201	

prices through efficient planning													
certain regulations are imposed by the government so that a transport company is only allowed to cover certain route kilometers												218	
Driving technology, I can do a bit with speed													88
across the entire supply chain													96
Speed limit													120
the ecological footprint is completely different if I drive evenly.													125
the costs of CO_2 emissions will increase										175			
a better density of goods being transported on individual routes										198			

Block F Technical sustainability	Interview partner (statements and line number)												
Statement - Reduction I	Experte 2	Experte 1	Experte 3	Experte 9	Experte 4	Experte 5	Experte 6	Publisher	Experte 8	Experte 10	Experte 11	Experte 12	Experte 13
Electric vehicles need to be charged much faster		115		215									
There is not much left to optimize with combustion engines		119											
Preferably only direct connections without having to change trains		136											
Through automation, digitalization and faster communication		142											

Abuse and attempts to penetrate there can increase massively	203												
Using AI instead of exerting yourself first. All this consumes energy, a lot of energy	207												
Electrify more	213												
electricity demand will increase sharply	215												
Share server and computing power	238												
the digital automatic coupling			440	201								126	
The rail employees are running out of shunting personnel			442										
If a rail company can no longer maneuver, I have to ask how it can secure its future			444										
A large number of wagon groups are also operated in CT			445										
Assistance systems in scheduling and control			491										
Only when everything is in the cloud can modernization begin			529										
The 2030s to 60s is the age of engineers to find solutions to actual baseline problems again					253								
Vacuum-based hyper-loop systems can be useful here					261								
Fuel cell, possibly the next thing					261								

There must be a common rail network in Europe					271								
Utilization planned manually; allocation of capacities planned manually					301								
PCs are often more like an archive or as a place of documentation than as support					303								
Of course, there is potential for optimizing network utilization						212							
extremely densely populated network, which is what we have in Germany in particular.						215							
AI requires resources because it consumes electricity first						218							
It will be an energy mix, not just a purely electric solution							231						
will always have a certain proportion of combustion engines							232						
Renewable drives will determine the topic of hydrogen and electricity							235						
decision is likely to fall to the sustainable solution rather than the environmentally damaging one.							263						
the reduction of battery costs								181					
The technology of the internal combustion engine is already far from								185					

exhausted anyway													
These are continuous small steps								189					
Cargo tube								189					
Away from the roads and onto the rails								192					
where two tracks are currently running, a third could be added								197					
Automation and digitalization are opportunities	201							210					
increase traffic density and maintain safety on the road	200							211					
Autonomous driving is making significant progress								222					
Introducing platooning for trucks							103	216					
Develop technologies to avoid having to transport goods in the first place									309				
Solution via 3D printing									312				
Only transport the base material									319				
Founding a 3D printing center									328				
Many empty runs are to be avoided									338				
AI can optimize travel routes									353				
Relevant technology that moves away from combustion engines				206									
Reducing empty runs through IT and AI				229									

Building networks between freight forwarders				231									
Trend away from private transport											152		
IT is exerting even more influence in all areas of transportation											155		
IT also controls traffic flows											159		
AI plays a major role and has an impact on traffic management in Europe											161		
digitize, establish smart cities, provide real-time information												231	
Increased use of sensors												236	
Optimize maintenance												240	
Derive measures that recognize a charge state, whether it has slipped												242	
the introduction of the ETCS standard can reduce headways												249	
Shortening construction times with AI												255	
Evaluate terrain topographies to identify problems at an early stage												260	
Logistics system sufficiently well networked												274	
Evaluation of camera images												287	
Automation, digitalization and flanked by a strategic apology													133

when danger arises, the brakes are applied immediately, nobody can react that quickly													136
More data is collected nowadays										228			
high capture of data streams. and on the other hand with stronger computing power										230			

Block G Green Energy	Interview partner (statements and line number)												
Statement - Reduction I	Experte 2	Experte 1	Experte 3	Experte 9	Experte 4	Experte 5	Experte 6	Publisher	Experte 8	Experte 10	Experte 11	Experte 12	Experte 13
Subsidy on the energy price	246	150			318								
Produce green energy yourself		151											
Use fines or punitive tariffs, also for fossil energy		154			314								
Infrastructure must be in place		163											
Everything must remain affordable		165											
More weight due to batteries		172											
Banks should increasingly pay attention to what is more sustainable	249												
Communication is often forgotten	256												
Facts should also be explained	257												
live with the elements we have	270												
Closing gaps in the network through electrification			561										
There are still relatively many diesel locomotives in			563										

the shunting area today													
The locomotives in the shunting area are getting on in years			572										
CT also needs pre- and post-carriage with trucks			611										
There is no holistic approach			624										
Using hybrid locomotives			628										
Only 60% of the route has overhead lines, a quota should be set			663										
Cost of capital due to high uncertainties					321								
The coal phase-out is already enshrined in law						225							
We are committed to promoting sustainability worldwide						237							
then you have to switch to low-CO_2 alternatives						247							
Monetary issue, only for a few people is being green worth more money							272						
a whole host of own barriers							275						
It is important to create a holistic understanding							279						
Disused railroad lines could be made usable							294						
Dealing more responsibly with greenhouse gas emissions							312						

Make conventional energy more expensive								233					
to reduce the existing subsidies for fossil fuels								236					
Ensure good communication								240					
not preach renunciation, but present the transport issue as an opportunity								241					
infrastructure that enables as much electrification as possible								255					
only hydrocarbon-based energy sources have the necessary energy density								261					
Subsidies and taxation									376				
Create dedicated infrastructure for freight transport									387				
Automated transshipment centers									400				
Promotion of green Driving technology									402				
Incentives via CO_2 levy				246									
Theoretical willingness to change				262									
Converting diesel locomotives to hybrid				266									
Provide infrastructure											180		
the range of electric cars											192		
Use of synthetic fuels											197		

as emission-free as possible, and to ensure energy consumption											217		
The monetary aspect is negative for companies												305	
Electrification, attempts to reduce diesel locomotives												329	
The sustainability aspect must take place in people												335	
Orderable robotaxis as an alternative												348	
Thinking of your own benefit												350	
greater reliance on battery electric trucks										243			
complement European cooperation										268			
greenhouse gases is viewed holistically										270			

Block H Exit	Interview partner (statements and line number)												
Statement - Reduction I	Experte 2	Experte 1	Experte 3	Experte 9	Experte 4	Experte 5	Experte 6	Publisher	Experte 8	Experte 10	Experte 11	Experte 12	Experte 13
Promoting sustainability and green electricity		183											
Offsetting greenhouse gases		184											
Reforesting forests		185											
Moses on the wall		186											
Simplifying cooperation between countries	279												
more networking	281												

Extreme focus on CO_2 neutrality						263							
Alternative fuels can save further emissions							319						
synthetic fuels such as e-fuels							322						
Availability of hydrogen and its derivatives								280					
Electrify as much as possible								289					
there is definitely not enough hydrogen								298					
Weight reduction for freight cars									410				
If a freight car is not so heavy, it consumes less energy									415				
It takes at least 7 years to approve a product									428				
Further expanding the network				280									
Expand transparency				284									
lead to people turning away from these ideas											247		
The buzzword is air cabs or drone logistics	xxx											360	
Combination with different means of transportation												174	
Infrastructure measures must be linked to the energy infrastructure										301			
Thinking of transformation even more holistically										308			

Table 20: Paraphrasing and reduction I
Source: Own representation.

In this table, the statements of the results are summarized and consolidated. This is done for each expert and by naming the lines of the respective transcript. This results in the reduction I.

Appendix 9: Generalization and reduction II of transcripts

Question B1: Where do you see the greatest potential for increasing sustainability?

Cat.	Line	Expert	Paraphrase	Generalization	Reduction II
C1	12	Experte 1	Potential lies in the collective load	The solution is professionalism	**C1.1** Shift to rail ▪ Groupage traffic ▪ Shift to rail ▪ Electrification ▪ Potential for sustainability issues is rated highly ▪ Potential not yet harnessed
C1	16	Experte 1	Shift traffic from road/airplane to rail	Shift to rail	
C1	43	Experte 4	The biggest lever in transport planning lies in electrification	Electrification is the best solution	
C1	65	Experte 5	Potential is not yet highly valued	Potential is not highly valued	
C1	88	Experte 8	also optimization possibilities in the process	Optimization through digitalization	**C1.2** Mobility ▪ Digitalization is seen as a driver ▪ DAC creates capacities ▪ Networking between market participants ▪ The energy generation itself ▪ Range of e-vehicles is not sufficient ▪ collect more data
C1	96	Experte 8	Automate DAC	DAC contributes to automation	
C1	34	Experte 9	Networking modes of transport	Networking	
C1	37	Experte 9	Combine transport with each other	Networking	
C1	13	Experte 13	In energy generation itself	Energy generation	
C1	192	Experte 11	Range of e-cars	Increased range	
C1	228	Experte 10	Much more data is collected nowadays	Greater data collection on mobility	
C1	105	Experte 8	a lot of energy only during acceleration and braking	Energy consumption during breaking and acceleration	**C1.3** Technology ▪ Braking and accelerating require energy ▪ Noise protection is also part of sustainability ▪ Making rail more attractive
C1	137	Experte 8	Green rail technology is climate,	Noise protection is also part of sustainability	

			nature and noise protection		▪ Creating transparency about activities ▪ Formation of alliances ▪ Renewing existing routes
C1	144	Experte 8	Quiet trucks or optimized wheelsets reduce noise	Noise protection thanks to modern trucks and wheelsets	
C1	282	Experte 2	Making rail more attractive	The rail is not in demand	
C1	55	Experte 11	Merger of alliances	Formation of alliances	
C1	69	Experte 11	Renewing existing routes	Renewal	
C1	116	Experte 3	freight transport its own network	Own network	**C1.4** Combined transport ▪ Freight transport needs its own network ▪ Automating and digitizing combined transport ▪ perceived as simple and good
C1	202	Experte 8	Optimize through digitization and automation	Digitization and automation	
C1	208	Experte 8	Piggyback truck	Solution: Combined transport	
C1	224	Experte 8	Piggyback's idea is simple, obvious and good	Solution: Combined transport	
Question B2: Where do you see the greatest risks in increasing sustainability?					
Cat.	**Line**	**Expert**	**Paraphrase**	**Generalization**	**Reduction II**
C1	77	Experte 6	Raw materials for batteries must be extracted from the earth	The extraction of raw materials leads to exploitation	**C1.5** Electrification ▪ The extraction of raw materials leads to environmental damage ▪ Sustainability must be considered in its entirety ▪ Hydrogen as an alternative Complete electrification is not possible ▪ It is difficult to operate an entire truck with a solar system
C1	78	Experte 6	Sustainability not considered far enough	Taking care of the big picture in sustainability	
C1	24	Experte 13	We get bogged down	Change technology slowly	
C1	52	Publisher	Import of hydrogen too slow	Hydrogen as an alternative	
C1	133	Experte 8	complete electrification of the grid will not happen	Complete electrification is not possible	

C1	44	Experte 1	It is difficult to operate a truck with solar energy	A truck cannot be powered entirely by solar energy	▪ There is little electrification in urban areas ▪ High prices burden transportation ▪ Electricity mix brings clarity ▪ Certificate purchasing is contra productive
C1	107	Experte 3	Rural areas hardly electrified	Electrification in the countryside	
C1	53	Experte 9	Infrastructure was neglected	Infrastructure neglected	
C1	31	Experte 11	Provision becomes expensive because it puts a huge strain on traffic	Prices burden transportation	
C1	69	Experte 9	Reduction of fossil fuels	Electricity mix	
C1	19	Experte 10	Electricity mix and no certificate purchase	Electricity mix	
C1	126	Publisher	who has the ambition to drive 1000 km non-stop	1000 km non-stop is not yet technically possible	**C1.6** Limiting the driving distance ▪ Trucks cannot drive 1,000 km non-stop ▪ Equivalent infrastructure for all modes of transport ▪ The trend is moving away from CT due to costs and duration ▪ Using energy correctly ▪ Overview of expansion topics
C1	60	Experte 9	Equivalent infrastructure for all modes of transport is lacking	Equivalent infrastructure	
C1	225	Experte 8	even if it takes longer for the train to get down there and it's expensive	Transportation by train takes longer and costs more	
C1	270	Experte 3	Massive loads of CT go onto the street	Shipments on the road	
C1	61	Experte 11	Using energy correctly	Using energy correctly	
C1	70	Experte 12	You can't keep up with the expansion	Keep an overview	

Question C1: How can European freight transport pay attention to environmental sustainability in the future?

Cat.	Line	Expert	Paraphrase	Generalization	Reduction II
C2	18	Experte 1	Rail is the most sustainable mode of transport	The rail is sustainable	**C2.1** Rail is sustainable ▪ Rail is the most sustainable mode of transp. ▪ Pay attention to emissions and what your tanker up with ▪ Hydrogen as an alternative ▪ Use optimal capacity utilization ▪ Reforesting forests ▪ Using biofuels
C2	58	Experte 3	Also pay attention to what you put in the tanker	Pay attention to what your tanker	
C2	38	Experte 6	Hydrogen is an alternative	Hydrogen as an alternative	
C2	50	Experte 4	generate optimum capacity utilization	Optimal utilization	
C2	185	Experte 1	Reforesting forests, greening walls	Reforesting nature	
C2	59	Experte 10	Bio-fuels	Using bio-fuels	
C2	289	Publisher	Electrify as much as possible	Electrification	**C2.2** Mobility ▪ Electrify as much as possible ▪ CT needs upstream and downstream ▪ Switching from terminals to renewable energy ▪ Using hybrid drives ▪ Creating a holistic view of all players ▪ Truck has high sustainability potential ▪ Away from diesel locomotives, towards electric mobility ▪ Supporting the use of synthetic fuels ▪ CO_2 neutral production ▪ Reduction of diesel locomotives through electrification
C2	601	Experte 3	get to 100% renewable energy from terminal to terminal	Renewable energies make sense	
C2	611	Experte 3	Combined transport also needs pre-carriage by truck	Pay attention to the leading and trailing edge	
C2	628	Experte 3	Using hybrid locomotives	Using hybrid locomotives	
C2	90	Experte 10	Use ecological technologies	Ecological technologies	
C2	93	Experte 10	View as a whole	Creating a uniform view	
C2	57	Experte 10	Road-based truck traffic, high sustainability potential	Truck has high sustainability potential	
C2	107	Experte 12	Getting away from diesel locomotives and more electric mobility	From diesel locomotives to electric mobility	

C2	197	Experte 11	Use of synthetic fuels	Energy source	
C2	217	Experte 11	As emission-free as possible	Energy source	
C2	329	Experte 12	Electrify, try to reduce the number of diesel locomotives	Electrification	
C2	61	Publisher	away from fossil fuels	Do not use combustion engines	**C2.3** Do not use combustion engines ▪ Fuel cells are clean and is perceived as an alternative ▪ Use alternative fuel carriers and also as a mix ▪ Even when switching to electric ▪ Defective batteries are a problem when it comes to disposal ▪ Landscapes destroyed ▪ Expansion of the overhead lines ▪ Routing via induction
C2	78	Publisher	Development in the fuel cell sector appears to be cautious	Fuel cell as an alternative drive system	
C2	30	Experte 2	Increased use of alternative fuels	Use alternative fuel carriers	
C2	66	Experte 2	going completely electric also requires somewhere to generate electricity	Electricity must be generated	
C2	80	Experte 2	What to do with batteries that need replacing	Used batteries must be disposed of	
C2	84	Experte 1	the A5 overhead line for trucks, this could be a solution	Operating e-trucks via overhead lines	
C2	86	Experte 1	Routing via induction or across the floor	Making e-trucks via induction or in the ground	
C2	88	Experte 3	Making the battery, drive and overhead line more sustainable.	Batteries and drives need to be made more sustainable	**C2.4** Construction of terminals ▪ Making batteries, drives and overhead lines more sustainable ▪ Sustainability must be affordable ▪ More terminals should be built
C2	91	Experte 3	Should then also be affordable	Sustainability must be affordable	
C2	132	Experte 3	Building more terminals for comb. transport	Building terminals	

C2	146	Experte 3	Railcars can also travel 20 kilometers without overhead lines	Driving on batteries for a short period of time	▪ Locomotives should travel short distances without overhead lines ▪ Pay more attention to the first and last mile in future ▪ Shorten the distances between vehicles ▪ Establish noise barriers
C2	170	Experte 3	Think more about the first and last mile in future	Paying attention to the last mile	
C2	95	Experte 6	Small distances mean fewer emissions	Short distances reduce drag	
C2	102	Experte 12	Noise barriers must be designed in such a way that they are beneficial to the environment	Establish noise barriers	

Question C2: What needs to be improved in the construction of infrastructure and the development of drives?

Cat.	Line	Page	Paraphrase	Generalization	Reduction II
C2	36	Experte 5	CO_2 Improve balance	Working on the CO_2 balance and the CO_2 footprint	**C2.5** Alternatives ▪ Pay attention to CO_2 balance and CO_2 footprint ▪ Shifting from air freight to rail saves emissions ▪ Green hydrogen ▪ Planting of buildings ▪ Plants absorb CO_2 and produce oxygen ▪ e-Fuels as an alternative ▪ The availability of hydrogen is not sufficient ▪ Hydrogen must be available in large quantities ▪ The energy mix is crucial ▪ Do not uproot plants ▪ Do not seal the soil
C2	48	Experte 6	You could save 90% of air freight emissions by going by rail	Shifting from air traffic to rail saves more than 90% of emissions	
C2	66	Experte 6	Appropriate infrastructure and the availability of hydrogen are a must	Green hydrogen as an alternative	
C2	322	Experte 6	synthetic fuels such as e-fuels	e-Fuels as an alternative	
C2	280	Publisher	Availability of hydrogen and its derivatives	Ensuring the availability of hydrogen	
C2	282	Publisher	Making hydrogen available globally in large quantities and also distributing it globally	Hydrogen must be available in large quantities	

C2	32	Experte 12	Renewable energies in the energy mix	Energy mix	▪ Define CO_2 footprint ▪ Explanation of why infrastructure measures are necessary for an improved CO_2 footprint
C2	30	Experte 13	planted especially for this purpose, but it was also cleared for it	It should be planted and not cleared	
C2	100	Experte 12	Do not seal the soil completely	Do not seal the soil	
C2	43	Experte 13	CO_2 -print that is clearly defined	Define CO_2 footprint	
C2	163	Experte 10	Explanation of why infrastructure measures are necessary	Explain why infrastructure measures are necessary	
C2	50	Experte 2	Cargo-Loop, these solutions could also be made underground	Fast transportation, underground if necessary	**C2.6** Cargo loop ▪ Cargo Loop and Hyper Loop are perceived as positive
C2	123	Experte 4	Research and development of drives	Research must continue on the drives	**C2.7** Development ▪ Research needs to be carried out on drivers ▪ e-Fuels contribute to sustainability ▪ Pipelines need to be built to supply hydrogen ▪ The financial outlay is immense ▪ Provide infrastructure for new drives and operating materials ▪ The fuel cell could be an alternative ▪ Freight transport is outdated and needs to be modernized
C2	130	Experte 4	e-Fuels contribute to ecological sustainability	e-Fuels as an alternative	
C2	83	Experte 5	Trucks that convert to e-trucks	Truck conversion	
C2	101	Experte 5	drives must be converted so that they are actually suitable for sustainable fuels such as methanol and hydrogen.	Convert drives	
C2	115	Experte 6	generate alternative drive systems	Generating alternative drive systems	
C2	128	Experte 6	To develop drives, you need the infrastructure	Ensuring the infrastructure	

C2	80	Publisher	Further research into fuel cell technology is needed	Fuel cells could be an alternative	▪ Building digital interlockings ▪ Converting transshipment points ▪ Automation of processes ▪ Introduce eco-certification
C2	154	Experte 8	Through automation	Increase automation	
C2	171	Experte 8	Digitizing freight transport	Driving digitalization forward	
C2	184	Experte 8	Freight cars are as they were 100 years agom	Freight cars and technology are outdated	
C2	193	Experte 8	Optimization of transshipment points	Optimization	
C2	175	Experte 12	Introducing an eco-certification	Introduce eco-certification	
Question D1: Which aspects will play a greater role in European freight transport than they do today?					
Cat.	**Line**	**Expert**	**Paraphrase**	**Generalization**	**Reduction II**
C3	36	Experte 3	People must be prepared to accept change	everyone has to embrace the changes	**C3.1** Rethinking ▪ Shippers, customers and employees need to rethink their approach ▪ When sustainability costs something, many people opt out ▪ RoI is too far in the guild ▪ Companies fear financial impact ▪ Sustainability as an opportunity ▪ Openness to technology ▪ Looking at the entire production chain ▪ Creating public consensus ▪ More control ▪ Create transparency ▪ Traceability
C3	77	Experte 4	RoI is a long way off	The RoI is too far away	
C3	88	Experte 4	Consume less and therefore produce less	Reducing consumption reduces production	
C3	58	Experte 5	Companies have concerns about the impact on finances	Companies fear financial impact	
C3	31	Experte 6	Sustainability is great as long as it doesn't cost anything	Sustainability must not cost anything	
C3	241	Publisher	not preach renunciation, but present the whole thing as an opportunity	Sustainability should be seen as an opportunity	
C3	94	Experte 9	Openness to technology	Openness to technology	

C3	95	Experte 9	Consideration of the production chain	Manufacturing chain	▪ Acceptance must be achieved ▪ Avoid greenwashing ▪ Away from private transport
C3	103	Experte 9	Creating public consensus	Creating public consensus	
C3	79	Experte 11	there must be more control	More control	
C3	88	Experte 11	Create transparency	Create transparency	
C3	76	Experte 13	Traceability recognizable	Traceability	
C3	79	Experte 13	Acceptance must be achieved	Acceptance must be achieved	
C3	145	Experte 12	Greenwashing should be avoided	Avoid greenwashing	
C3	58	Experte 1	Truck drivers must be able to spend a reasonable night on the road	Drivers must be able to spend the night properly	**C3.2** Traveling personnel ▪ Create overnight accommodation for drivers ▪ Drivers must have enough money with them Rooms must be booked in advance by the company ▪ Pay attention to the use of subcontractors ▪ Comply with the minimum wage ▪ Ensure compliance with ethical principles ▪ Social standards must rise, salaries often does not match the job ▪ Nobody wants to do manual and heavy work anymore
C3	214	Experte 3	Drivers have no money to pay for an overnight stay	Drivers often have no money for overnight accommodation	
C3	236	Experte 3	Multi-level subcontractor contracts are in place	Pay attention to subcontractors	
C3	159	Experte 4	Compliance with ethical principles	Ethical principles	
C3	133	Experte 5	Social standards must be raised even further	Social standards must increase	
C3	54	Experte 1	Drivers' salaries are not sufficient	Salary does not match task	
C3	255	Experte 3	Foreign companies still pay less than the minimum wage here	Compliance with the minimum wage must be ensured	

C3	157	Experte 8	Walk the entire train and check every brake on every freight car	Nobody wants to do manual vehicle inspections anymore	▪ Autonomous driving requires training and instruction ▪ Wage dumping must be avoided ▪ Pay attention to hourly wages and working conditions ▪ Create more VDU workstations ▪ Protect employees from hazardous substances and cleaning agents
C3	117	Experte 5	Autonomous driving	Autonomous driving	
C3	149	Experte 9	Creating awareness about hourly wages and working conditions	Hourly wages and working conditions	
C3	128	Experte 12	Employees are relieved of the heavy burden	Relieving employees of heavy workloads	
C3	131	Experte 12	Pay attention to hazardous substances	Protect employees	
C3	132	Experte 12	Pay attention to cleaning products	Protect employees	
C3	146	Experte 12	On-site workstation becomes a VDU workstation	Computer workstations	
C3	121	Experte 5	Fluctuation is very high	Fluctuation is very high	**C3.3** Competitiveness ▪ Fluctuation is very high ▪ Through subsidy ▪ Through promotion ▪ eliminates jobs that nobody wants to do ▪ creates pleasant jobs ▪ Automation in the shunting environment relieves train drivers ▪ Is carried out on the backs of low-wage countries
C3	143	Experte 5	especially state subsidies	Subsidy and promotion	
C3	149	Experte 5	Waiving taxes or support is the biggest lever	Taxes are a lever	
C3	265	Experte 8	Automation eliminates jobs that nobody wants to do	Automation eliminates jobs that nobody wants to do	
C3	270	Experte 8	Automation creates jobs that are pleasant for people	Driving automation forward	
C3	272	Experte 8	Nobody wants to be a train driver anymore	Nobody wants to make a log guide	

C3	121	Experte 9	On the backs of low-wage countries	Low-wage countries	▪ Import of labor and distortion of competition ▪ Fair salary and benefits structure
C3	124	Experte 9	Import of labor and thus distortion of competition	Distortion of competition	
C3	78	Experte 11	Payment and social benefits for drivers must be ensured	Fair salary and social benefits	
C3	274	Experte 8	Automation and remote control	Driving automation	**C3.4** Mobility ▪ Autonomous driving ▪ Shifting from road to rail reduces noise ▪ Making trucks quieter ▪ Making transshipment points quieter ▪ Optimize logistics planning ▪ Ensuring justice ▪ Speed limit in traffic jams ▪ Full loading leads to better utilization and density
C3	286	Experte 8	Trucks and transshipment points produce noise	Noise must be reduced	
C3	192	Publisher	Away from the roads and onto the rails	Shift from road to rail	
C3	85	Experte 10	Optimize logistics planning and avoid empty runs	Optimize logistics planning	
C3	92	Experte 11	Put a stop to injustice	Ensuring justice	
C3	120	Experte 13	Introduction of speed limits	Speed limit in traffic jams	
C3	198	Experte 10	Better density of goods to be transported	Utilization of loading equipment	

Question D2: What measures and incentives could there be to increase social acceptance?

Cat.	Line	Expert	Paraphrase	Generalization	Reduction II
C3	213	Experte 3	Building more truck stops	Building more and larger truck stops	**C3.5** Expansion ▪ Expansion of truck stops ▪ The expansion of infrastructure costs money, and this money must be secured ▪ Buying green transportation
C3	284	Experte 3	Funds are made available for the expansion of the infrastructure	Funding must be secured	
C3	160	Experte 6	Buying green transportation solutions	Green transportation must be purchased	

C3	299	Experte 8	DAC to promote this	Promote DAC	▪ The introduction of the DAC would be enormous help ▪ Benefits for environmentally friendly technologies ▪ Progressive CO_2 taxation ▪ Willingness to change ▪ Grid expansion ▪ Creating transparency
C3	294	Experte 8	Progressive CO_2 taxation	Taxation	
C3	296	Experte 8	Tax concessions for environmentally friendly technologies	Tax concessions	
C3	262	Experte 9	Willingness to change	Willingness to change	
C3	280	Experte 9	Further expanding the network	Grid expansion	
C3	284	Experte 9	Expand transparency	Transparency	

Question E1: How can freight transport ensure economic sustainability?

Cat.	Line	Expert	Paraphrase	Generalization	Reduction II
C4	67	Experte 3	The economy should use resources wisely	Improved use of resources	**C4.1** Company image ▪ Improved use of resources ▪ Companies shy away from investments ▪ Companies pay attention to their image ▪ A train that rolls once needs very little energy ▪ Availability of wagons ▪ Availability of loading equipment ▪ Combined transport increases capacities ▪ Pay attention to track access charges ▪ Pay attention to capacity utilization ▪ Quality must be increased ▪ Combined transport is
C4	60	Experte 5	Companies shy away from investments	Investments as a brake	
C4	64	Experte 5	Companies have reputational risk	Pay attention to the image	
C4	175	Experte 2	There will be no reduction in rail freight, we need to get much more freight onto the rails	More freight needs to be transported by rail	
C4	180	Experte 2	Sufficient wagons must be available	Provide availability of loading equipment	
C4	318	Experte 3	Truck companies can use the CT and thus have explosive capacities	Combined transport to increase capacities	
C4	261	Experte 3	Combined transport is sold as the great savior	Combined transport as an alternative	

C4	231	Experte 9	Building a network between freight forwarders	Communication and collaboration	seen as the savior ▪ Communication and cooperation between market participants ▪ Using sustainable materials ▪ Acting more transparently ▪ Increasing attractiveness through station planning ▪ Ensure efficient planning ▪ Prices will rise due to costs for CO_2 levies ▪ Use of sharing concepts
C4	93	Experte 12	Increasing attractiveness in station planning	Increasing attractiveness through station planning	
C4	98	Experte 12	Material selection is a major aspect of sustainability	Materials are important	
C4	201	Experte 12	Efficient planning for competitive and attractive prices	Efficient planning	
C4	175	Experte 10	That the cost of CO_2 emissions will increase	Countering cost increases	
C4	136	Experte 13	AI reacts faster to danger than a human could	AI reacts quickly	
C4	200	Experte 10	Using sharing concepts	Sharing concepts	
C4	158	Experte 2	Tax concessions are one solution	Tax concessions are an incentive	**C4.2** Financing ▪ Incentive Tax benefits ▪ Increase road tolls to avoid unnecessary driving ▪ Financing risk issues is difficult to quantify ▪ Solution financially negative for now
C4	105	Experte 1	Increase road toll	Increase road toll	
C4	106	Experte 1	use a portion of the fuel tax	Making sensible use of taxes	
C4	47	Experte 12	Financing risk issues is difficult to quantify	Financing of risk issues	
C4	305	Experte 12	monetary aspect is initially negative	Financially unattractive at first	
C4	203	Experte 4	Logistics is the backbone of a modern society	Logistics is the backbone of society	**C4.3** Importance of logistics ▪ Logistics is the backbone of society ▪ Without transport logistics, there would be no deliveries
C4	206	Experte 4	that without transportation and logistics, not that much actually works	Transport logistics ensure supplies	

C4	163	Experte 8	Optimize downhill and braking control to save energy	Start and break as little as possible	▪ Logistics ensures prosperity ▪ Vehicle fleets cannot be completely converted ▪ Development of charging infrastructure ▪ Changing demand behavior ▪ Laying the foundations ▪ Re-harnessing the locomotives ▪ Forming alliances ▪ Promoting alliances ▪ Mapping traffic flows ▪ with CO_2 tax and make prices more transparent ▪ The last mile is important
C4	235	Experte 6	New drives will determine the topic of hydrogen and electricity	New drives decide on the fuel	
C4	239	Experte 6	Water flow infrastructure	Infrastructure development	
C4	72	Experte 12	Create the basis for realizing freight transport	Laying the foundations	
C4	76	Experte 12	Problems with the locomotives	Re-harnessing the locomotives	
C4	106	Experte 11	Promoting more alliances	Promote alliance	
C4	107	Experte 11	Mapping long traffic flows	Mapping traffic flows	
C4	110	Experte 11	Forming alliances to offer more transport connections	Forming alliances	
C4	192	Experte 12	The last mile will always be difficult	Last mile is important	

Question E2: The load capacity limit of the infrastructure has already been reached. How economic measures could relieve the burden on infrastructure in the future?

Cat.	Line	Expert	Paraphrase	Generalization	Reduction II
C4	45	Experte 3	to use the available resources of nature more consciously	Better stewardship of nature and the environment	**C4.4** Relieving the burden on infrastructure ▪ Reduction of unnecessary journeys ▪ Introduction of the road toll for passenger cars ▪ Sustainability must not cost anything ▪ Sensible mix of rail and road ▪ Inland vessel can supplement
C4	149	Experte 2	But when it becomes your own porte-monnaie and the goods become more expensive, the sustainability stops	Sustainability must not cost more	
C4	330	Experte 3	The optimal solution is a clever	Sensible mix between modes of transport	

			mix between truck and rail		▪ European Train Control System to be optimized ▪ Digitization of the signal boxes ▪ The state and politicians must take action ▪ Awareness for a better approach to nature ▪ Optimization of the route ▪ Making infrastructure more transparent ▪ Transparent presentation of the service life of the infrastructure ▪ Forming more alliances to offer more transport connections ▪ Avoid permanent construction sites ▪ Shift from fossil fuels to renewable energy sources
C4	345	Experte 3	Where there is plenty of time, the inland waterway vessel can be used	Inland vessel can also deliver capacity	
C4	410	Experte 3	Promoting digital interlocking systems	Digitize signal boxes	
C4	172	Experte 9	Create basic infrastructure	Create infrastructure	
C4	186	Experte 9	Optimization through route utilization	Optimization of the route	
C4	66	Experte 12	Infrastructure and connection of the route must be realized	Making infrastructure more transparent	
C4	107	Experte 10	The service life of the infrastructure must be presented transparently before it is financed	Transparent presentation of the service life of the infrastructure	
C4	110	Experte 11	Forming alliances to offer more transport connections	Offering more transport connections through alliances	
C4	121	Experte 11	Relief on the roads	Relief on local roads	
C4	131	Experte 11	Avoid permanent construction site	Avoid permanent construction sites	
C4	316	Experte 12	The shift from fossil fuels to renewable energy sources	From fossil fuels to renewable energy sources	
C4	161	Experte 5	cover the costs of their suppliers and increase prices to their customers	Companies pay attention to their contribution margins and profitability	**C4.5** Economic efficiency ▪ Companies pay attention to profitability and

C4	23	Experte 1	The infrastructure is outdated and does not meet the requirements	Outdated infrastructure needs to be modernized	contribution margins ▪ Outdated infrastructure needs to be modernized ▪ Efficient use of infrastructure ▪ Optimize speeds ▪ Regulation via variable fees
C4	444	Experte 8	Make existing infrastructure more efficient through automation	Efficient use of existing infrastructure	
C4	448	Experte 8	drive one behind the other at the optimum speed	Optimize speeds	
C4	187	Experte 9	Regulation via variable fees	Variable fees	
C4	163	Experte 5	everyone should continue to be supplied with everything, whether by rail or by truck	Supply by rail and road	**C4.6** Mobility ▪ Supply of markets by rail and road ▪ Reduction of the space occupied by trucks ▪ Do not reduce the trucks ▪ Development of larger trucks, gigaliners, lift transports
C4	208	Experte 6	The trucks and drivers should not be reduced, but the space should be condensed	Reducing the space occupied by trucks, not reducing trucks	
C4	209	Experte 6	Larger trucks such as gigaliners or more hub transports	Development and addition of trucks	

Question F1: How could technology change in the next few years?

Cat.	Line	Expert	Paraphrase	Generalization	Reduction II
C5	311	Experte 8	technologies that goods no longer need to be transported	3D printing makes it possible to print on site	**C5.1** 3D printing ▪ 3D printing allows a lot to be printed on site ▪ Many trips are canceled ▪ Establishment of print centers ▪ Transporting base material
C5	314	Experte 8	3D printing is a big topic - 3D Druckcenter	3D printing can be used to print	
C5	321	Experte 8	transport only the basic material	3D printing makes it possible to print on site	

C5	25	Experte 1	Freight transport should work more digitally	Driving digitalization forward	**C5.2** Technical changes ▪ Production networking ▪ Digitization ▪ Development of green fuels ▪ Weight reduction through new composite materials ▪ More payload due to lower tare weight ▪ Bringing wagons and locomotives up to date ▪ Driver assistance systems ▪ Charge batteries faster ▪ Hardly any optimization is possible with combustion engines ▪ Develop and use smaller transport containers ▪ Direct connections without relocation ▪ Optimization of processes in the transport chains ▪ Common rail network in Europe ▪ Too few charging points ▪ Conversion of locomotives to hybrid ▪ Technology for combustion engines is exhausted ▪ Update infrastructure
C5	281	Experte 2	More networking	More networking	
C5	55	Experte 4	Aerodynamics of them is still to be optimized	Making vehicles more aerodynamic	
C6	61	Experte 4	Switching to green fuels	Developing green fuels	
C5	415	Experte 8	Weight reduction, especially for freight cars	More can be loaded due to weight reduction	
C5	419	Experte 8	If a freight car is not so heavy, it requires less energy	More can be loaded due to weight reduction	
C5	421	Experte 8	The lighter a wagon is, the more can be loaded onto it	More can be loaded due to weight reduction	
C5	100	Experte 3	Condition like the 60s	Wagons obsolete	
C5	244	Experte 8	improve battery technology	Battery charging technology	
C5	127	Experte 1	Use smaller transport containers that are then pulled on rails	Use smaller transport containers	
C5	136	Experte 1	Only direct connections are best	Direct connections without relocation	
C5	271	Experte 4	A common rail network must be created in Europe	Creating a common rail network	
C5	70	Experte 1	Create charging points for e-vehicles	There are too few charging points	
C5	185	Publisher	The technology of the internal combustion engine is already far from exhausted anyway	Technology for combustion engines is exhausted	

C5	206	Experte 9	Relevant technology that moves away from combustion engines	Away from combustion engines	
C5	266	Experte 9	Converting diesel locomotives to hybrid locomotives	Conversion to hybrid locomotives	
C5	67	Experte 11	Bringing the entire infrastructure up to date	Update infrastructure	
C5	142	Experte 1	Through automation, digitalization and faster communication	Automation and digitalization lead to success	**C5.3** IT + digitization ▪ Automation and digitalization lead to success ▪ Fast communication is the key to success ▪ Introduce DAC ▪ Driver assistance systems ▪ System solutions need to be modernized ▪ Reduce empty runs ▪ Traffic flow control ▪ Increased use of sensors ▪ Automation
C5	238	Experte 2	Share server and computing power	Share server performance	
C5	440	Experte 3	Introducing digital automatic coupling across the board	Introduce DAC across the board	
C5	491	Experte 3	Assistance systems in scheduling and control	Introducing further driver assistance systems	
C5	79	Experte 1	Charging time of 2-3 h too long	Charging time is far too long	
C5	229	Experte 9	Reduce empty transport with AI	Reduce empty runs	
C5	159	Experte 11	IT also controls traffic flows	Control through IT	
C5	133	Experte 13	Automation and digitalization through strategic decisions	Automation and digitalization	
C5	236	Experte 12	Increased use of sensors	Sensors	
C5	180	Publisher	Fuel cells and batteries as an alternative	Fuel cells are an alternative	**C5.4** Optimization potential ▪ Fuel cell as an alternative ▪ Battery-powered drives as an alternative
C5	181	Publisher	the reduction of battery costs	Batteries are still too expensive	

C5	197	Publisher	where two tracks are currently running, a third could be added	Expansion of the rail network	▪ Costs for batteries too high ▪ Expanding the rail network, one rail is not enough ▪ Returning brake energy to the grid ▪ Replace all combustion engines ▪ Increase traffic density to be efficient ▪ Introducing platooning for trucks ▪ Avoid educational trips
C5	211	Publisher	increase traffic density maintain safety	Traffic density must be increased	
C5	59	Experte 8	Modern trains can feed braking energy back into the grid	Feed energy back into the grid	
C5	78	Experte 8	Diesel vehicles are not scrapped for financial reasons	Replacing combustion engines with vehicles is not affordable	
C5	216	Publisher	Platooning for trucks	Platooning	
C5	340	Experte 8	Avoid many parallel trips	Avoid empty runs	

Question Q2: How can IT and AI solutions increase sustainability?

Cat.	Line	Expert	Paraphrase	Generalization	Reduction II
C5	26	Experte 1	Accelerate the exchange of information	Sharing information is the key	**C5.5** Exchange of information ▪ Accelerating the exchange of information between stakeholders ▪ PC for documentation/archive ▪ More collaboration ▪ Networking of systems
C5	303	Experte 4	PCs are often as an archive or as a place for documentation as support	Computers and IT serve as an archive	
C5	287	Experte 12	Evaluation of camera images	Evaluation of data	
C5	52	Experte 4	Companies are interested in investing in digital solutions	Investing in digital solutions	**C5.6** AI solutions ▪ Companies want to invest in digital solutions ▪ Break up single-track routes and lay on another track ▪ General refurbishment of the rail network
C5	101	Experte 3	Trains have to share a track	Eliminate single-track routes	
C5	128	Experte 3	Infrastructure needs to be completely renovated	Renovation of the infrastructure	

C5	326	Experte 8	In rail transport, AI can be used to optimize the entire travel chain	AI can optimize the travel route	▪ AI optimizes travel routes ▪ AI is used to evaluate traffic data ▪ Reduce construction times with AI ▪ Control terrain and vegetation by AI ▪ AI is used for monitoring ▪ AI can minimize the distances between trains ▪ Maintenance optimization
C5	218	Experte 5	AI requires more resources because it consumes electricity first.	AI requires the use of resources	
C5	161	Experte 11	AI has an influence on transport growth	AI is used for evaluation	
C5	240	Experte 12	Optimize and digitalize maintenance	Optimization of maintenance through AI	
C5	242	Experte 12	Measures in the event of a slipped load	AI is used for monitoring	
C5	255	Experte 12	Reduce construction times with AI	AI is used for evaluation	
C5	260	Experte 12	Evaluate topographies of	AI is used for evaluation	
Question G1: What incentives could politicians create to ensure that green energy is used in future?					
Cat.	**Line**	**Expert**	**Paraphrase**	**Generalization**	**Reduction II**
C6	96	Experte 13	Mapping across the entire supply chain	Control of the supply chain	**C6.1** Control system ▪ It takes too long until a product is ready for the market ▪ Subsidy as motivation ▪ Produce your own energy ▪ Subsidy or taxation ▪ Taxing fossil fuels
C6	427	Experte 8	European regulations are too high. You need at least 7 years	It takes several years to get to market maturity	
C6	151	Experte 1	Produce green energy yourself	Producing green energy	
C6	154	Experte 1	Use fines for the use of fossil fuels	Penalties for the use of fossil fuels	
C6	379	Experte 8	Through subsidies and taxation	Create monetary incentives	
C6	53	Experte 8	Increase energy efficiency	Increase efficiency	**C6.2** Increase efficiency

C6	163	Experte 1	Infrastructure must be in place	Infrastructure is necessary	▪ Infrastructure must be available and affordable ▪ Banks should promote green production ▪ Closing gaps in the network ▪ Incentives via CO_2 levy ▪ Cooperation with politics makes sense
C6	165	Experte 1	Everything must remain affordable	Sustainability must remain affordable	
C6	561	Experte 3	Closing gaps through electrification	Closing gaps in the network	
C6	246	Experte 9	Incentives via CO_2 levy	Incentives via CO_2 levy	
C6	177	Experte 9	The state must enable infrastructure competition	Promoting infrastructure competition	
C6	231	Experte 6	It will be an energy mix, not just a purely electric solution	A mix of energy is the solution	**C6.3** Raw materials ▪ Reduce fossil fuels ▪ Energy mix
C6	236	Publisher	to reduce the existing subsidies for fossil fuels	Reduce the promotion of fossil fuels	

Question G2: How do infrastructure, technology and drives need to change in order to be ready for future generations produce less greenhouse gas?

Cat.	Line	Expert	Paraphrase	Generalization	Reduction II
C6	225	Experte 5	The coal phase-out is enshrined in law	Legal regulation for the coal phase-out	**C6.4** Rethinking ▪ Legal regulations or resolutions, e.g. coal phase-out ▪ Communication with shippers, customers and the public ▪ Everyone needs to be aware of these issues ▪ Transparent sustainability ▪ Reduce certificate purchases
C6	256	Experte 2	Communication is forgotten	Optimize communication	
C6	247	Experte 5	raise awareness of the topic	Creating awareness of sustainability	
C6	55	Experte 12	Sustainability must first be transparent	Making sustainability transparent	
C6	224	Experte 11	Reduce selfish thoughts	Egoism	
C6	335	Experte 12	The sustainability aspect must be in people's minds	Rethinking	
C6	617	Experte 3	no charging infrastructure for battery-powered	Holistic view of combined transport	**C6.5** Drives and energy sources

			trucks traveling in CT?		▪ Charging units for batteries ▪ Choosing the right alternatives ▪ Promotion of green drive technology ▪ Increasing efficiency ▪ More electrification ▪ carbon-based energy sources have the necessary energy density ▪ e-Fuels can only be used to a limited extent ▪ Comparison of dirty and green electricity
C6	237	Experte 5	you have to switch completely to low-CO_2 alternatives	Switch to low-CO_2 alternatives	
C6	405	Experte 8	Promotion of green drive technology	Promoting green drive technology	
C6	250	Publisher	Drive types with energy sources that are as carbon-free as possible	Use of carbon-free energy sources	
C6	261	Publisher	Hydrocarbon-based energy sources have the necessary density	Hydrogen does not have the necessary energy density	
C6	157	Experte 6	Green and dirty electricity must be contrasted	Comparison of dirty and green electricity	

Question H1: Other points that have been forgotten?

Cat.	Line	Expert	Paraphrase	Generalization	Reduction II
C7	71	Experte 2	Holistic view of all participants	Holistic view of the modes of transport	**C7.1** Cooperation within the EU ▪ Improving cooperation between countries ▪ Consideration of the entire supply chain is important ▪ Establish ship-side hydrogen import structure ▪ Generate use cases and benefits for companies ▪ Society must adapt ▪ More engineering capacity will be needed in the next 30 years
C7	279	Experte 2	that cooperation between countries becomes easier	Political cooperation in the EU must be increased	
C7	109	Experte 2	Our society must change	Society must adapt	
C7	442	Experte 3	The rail employees are running out of shunting personnel	Rail companies must take care of additional shunting personnel	
C7	52	Experte 8	We could be much greener	increase green energy	
C7	294	Experte 6	Disused railroad lines could be made usable	Reactivating disused routes	

C8	153	Experte 4	Age of engineers to reinvent solutions such as the engine	Increase in engineering performance
C7	390	Experte 8	Dedicated infrastructure for goods	Dedicated infrastructure

Table 21: Generalization and reduction II
Source: Own representation.

Appendix 10: 4 x 3 matrix

Journals (secondary research)

Alexander Eisenkopf
Title: Mobility on the test rig
Year: 2023
Journal: Springer Publishing
Lnk: https://link.springer.com/article/10.1007/s10273-022-3170-z

Paula Quentin et al.
Title: Transport planning beyond infrastral change
Year: 2023
Journal: MDPI Publishing
Link: 10.3390/su151310025

Dadiana Dabija & Carmen Natasse
Title: Exploring the evolution of sustainability paradigm
Year: 2024
Journal: Research Gate
Link: 10.2478/picbe-2024-0066

Expertise (primary research)

Marius Dellbbrügge
Team Lead Transport Modeling and Process Planning - Institute of Transport Logistics.

Dr. Simon Experte 7
Senior Project Manager Hydrogen, Production Planning, Expertise in the field of Transport Logistics and Hydrogen.

Uwe Clausen
Director, Institute Trans-port Logistics, Tech. Uni Dortmund, Head of Fraunhofer Institute, author

Literature (secondary research)

David Grant et al.
Sustainable Logistics and Supply Chain Management: Principles and Practices for Sustainable Operations and Management, Kogan Page Publishing, 2022

Uwe Clausen
Advances in Resilient and Sustainable Transport, Cases studies, Springer Publishing, 2023

Elmar Bräkling
Logistics management with logistics power, fast and lean delivery, Springer Publishing, 2020

Chairs (secondary research)

tu technische universität dortmund

University: Technical University
Institute: Transport and Logistics
Owner: Prof. Dr. Uwe Clausen
Location: Dortmund

TUM Technische Universität München

University: Technical University
Institute: Logistics Technology
Owner: Prof. Dr. Johannes Fottner
Location: Munich

MIT Massachusetts Institute of Technology

University: Institute of Technology
Institute: Conveying Technology
Owner: Prof. Dr. xx xxx
Location: Cambridge Massachusetts

Figure 20: 4 x 3 matrix
Source: Own representation.

Appendix 11: Literature search table

No	Search term	Google Scholar	Ebsco and Research Gate	Own Book	Library	Springer Link	Google
1	CargoTransport volume						06.7.
2	Procurement logistics	09.07.	09.07.	28.06.	23.01.	12.07.	
3	Cargo-tubes			28.06.			
4	Digital sustainability	11.07.			23.01.	11.07.	02.07.
5	Energy technologies						02.07.
6	global warming	09.07.	09.07.				09.07.
7	Renewable energies						08.07.
8	European Commission	22.06.	22.07.				06.07.
9	European Green Deal	22.06.	22.07.				06.07.
10	Green Energy	22.06.	22.07.		23.01.		06.07.
11	Logistics basics			28.06.	23.01.		
12	Green logistics	21.06.			23.01.		03.07.
13	Cargo in Germany						03.07.
14	Cargo in Europe	10.07.				05.07.	10.06.
15	hyperloop systems	21.06.		21.06.	23.01.		21.06.
16	Issue analysis			10.07.			
17	Climate Protection Act						04.07.
18	Climate + targets	19.06.				05.07.	01.07.
19	Logistics management			14.07.	23.01.		
20	Mobility	03.07.	03.07.				03.07.
21	Mobility and logistics	27.06.	27.07.		23.01.		18.07.
22	Mobility concept	27.06.	27.07.				12.07.
23	Mobility competition	27.06.	27.07.		23.01.		03.07.
24	Sustainable develop	27.06.	27.07.				22.06.
25	Sustainable mobility	27.06.	27.06.			04.07.	10.06.
26	Paris climate targets						10.06.
27	Role of transport sector			18.06.			18.06.
28	Pillars of sustainability	28.06.	28.05.				11.07.
29	Value chain			18.06.	23.01.		18.06.
30	Technical sustainability	22.06.	22.06.				10.07.
31	Transportation net	11.07.	11.07.	15.06.	23.01.		04.07.

32	Greenhouse gases						04.07.
33	Greenhouse gas						04.07.
34	Trends in transport	09.07.	09.07.				20.07.
35	Switching to rail			08.07.			20.07.
36	Traffic in figures						22.07.
37	Decarbonization	18.06.	18.06.				20.07.
38	Transport destinations	19.06.	19.06.				19.07.
39							
40							
41							
42							
43							
44							
45							
46							
47							
48							
49							
50							

Table 22: Literature search table
Source: Own illustration - Search activities in 2025.

Appendix 12: PRISMA Flow Diagram

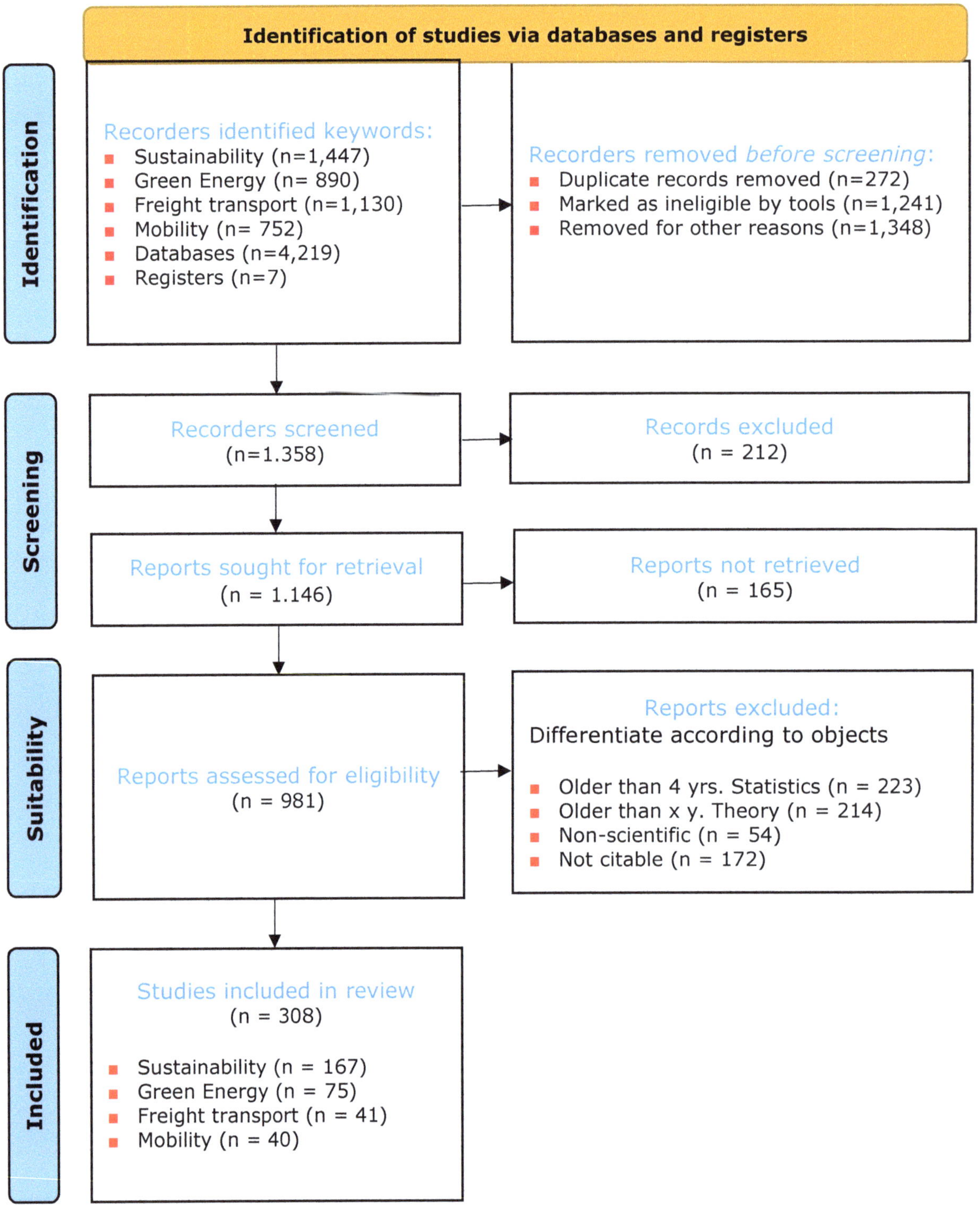

Figure 21: PRISMA Flow diagram
Source: Own illustration based on Page 2021.

The PRISMA flow diagram visually summarises the screening process. It initially records the number of articles found and then makes the selection process transparent by reporting on decisions made at various stages of the systematic review.

Appendix 13: Model research table

Source	Purpose	Type of Article	Reliability	Major Findings
Accenture *2024*	Climate effects of digitalization	Home Page	high	Study to assess the contribution of digital technologies to climate protection in Germany
Adams *2019*	Green freight transportation	Journal	very high	Impact of the supply chain
Alekseev *2024*	Assessment of the modern climate policy	Home Page	high	*Review of Business and Economics Studies*
Association of municipal enterprises *2024*	Effects of the EU climate target	Home Page	high	Possible impact of an EU climate target of -90% emissions for 2040 on Germany
Bagwan *2023*	Refrigeration technologies in refrigerated supply chains	Printed Book	very high	Suitability and sustainability of dry ice as a cooling medium
Baloise *2024*	Definition of mobility	Home Page	high	What do we mean by mobility
Banister *2022*	Unsustainable transportation	Home Page	high	City transport in the new century
Banister *2022*	Sustainable transportation	Home Page	high	Planning for cycling in urban environments
BDE *2023*	European Commission	Government	highest	European Commission must decide in 2026 on the possible inclusion of thermal waste recovery in the EU ETS from 2028
Behrends *2022*	Transportation and Lotistics Planning	Home Page	high	Transportation and Lotistics Planning
Black *2023*	The green logistics handbook	Printed Book	very high	Environmentally friendly solutions help to reduce costs and environmental impact
Blauwens *2022*	Transportation	Printed Book	very high	They therefore offer an economically and ecologically efficient solution for freight transportation

BMDV *2023*	The future needs paths	Govern-ment	highest	The share of total transport volume has remained constant at around 18-19% for many years
BMWK *2024a*	Reneawable Energy	Govern-ment	highest	Terms such as green or clean, renewable energy are used to describe energy sources that reduce the need for fossil fuels
BMWK *2024b*	ELEKTRA	Govern-ment	highest	ELEKTRA - Hydrogen for waterways
Bothe *2024*	The Energy Transition in Everyday Understanding	Home Page	high	Energy Transition in Everyday Understanding
Bouchery *2017*	Sustainable supply chain	Printed Book	very high	A Research-Based Textbook on Operations and Strategy
Bräkling *2020*	Logistics management	Printed Book	very high	Fast, lean and error-free delivery with logistics power
Brecha *2023*	Decarbonizing road freight transport	Journal	very high	The role of zero emission technologies
Browne *2023a*	Decarbonizing	Printed Book	very high	Decarbonization is the decisive key
Browne *2023b*	Urban freight	Printed Book	very high	Integrating Sustainability into Urban Freight Transport
Brunner *2020*	Impact of emerging transport	Research	high	Impact of emerging transport technologies on freight economic and environmental performance
BSL *2020*	Combined transport	Home Page	high	Importance and use of combined transport
Bundesklimaschutzgesetz *2019*	Climate Protection Act	Govern-ment	highest	
Business.gov *2025*	Zero-emssions	Home Page	high	Zero-emission zones to be introduced in many cities in 2025
BVL *2024*	Logistics objectives	Home Page	high	Digitalization and networking have increased the challenges for efficient material flow

Caldeirinha et al. *2023*	Role of cargo owner in Logistic chain sustainability	Journal	very high	
Calderinha *2023*	Role of cargo owner	Journal	very high	Role of cargo owner in logistic chain sustainability
Carter *2023*	Climate Change and maritime transport	Printed Book	very high	Adapting to climate change is a key challenge for the entire transportation industry
CISL *2025*	Sustainability in water unprecedated attack	Homepage	high	How to fight back. Creation of strategies
Climate Portal *2024*	Reneawable Energy	Home Page	high	Energy is generated by sun and wind
CM Logistics *2024*	Freight transport: rail versus truck	Home Page	high	The rail network is a huge construction site that drags on forever
Cole *2023*	Eco-friendly transport logistics	Printed Book	very high	Environmental Impacts of Transportation
Consultport. *2024*	7S Modell nach McKinsey	Home Page	high	
Courtier *2021*	Sustainable mobility and climate protection	Home Page	high	
Courtier *2023*	Green logistics	Printed Book	very high	Approaches to improving the environmental balance in freight transport
Coyle *2023*	Transportation	Printed Book	very high	Cargo Tubes offer a modern solution for optimizing supply chains
Crainic *2023*	Intermodal transportation	Printed Book	very high	By reducing environmental impact, cargo tubes contribute to achieving global climate targets
Crawford *2024*	The green transition	Homepage	high	

CT4EU *2022*	Intelligent networking	Home Page	high	Through the intelligent networking of freight transport contributes to a sustainable reduction in greenhouse gas emissions of up to 90 % and a reduction in energy requirements of up to 70 %.
Cullinane *2024*	Innovations in green logistics	Printed Book	very high	AI and telematics play a key role in mobility
Curtis *2024*	Handbook of sustainable Transport	Home-page	high	Transportation strategies must adapt to local conditions
Dabia & Nastase *2024*	Sylvicultura aeconomica	Home Page	high	The idea of sustainability dates back to the 17th century
Dabija & Nastase *2024*	Exploring the evolution of sustainability paradigms	Journal	very high	Linear circular economy models
Dalrada *2024*	Climate+ Technology, Clean Energy	Home Page	high	The use of renewable power does not threaten habitat loss
Davis et al. *2024*	Smart automated solution for public transport system	Home Page	high	*10th International Conference on Advanced Computing and Communication Systems*
De Feo *2019*	Economic transportation	Journal	very high	Economic, energy and environmental efficiency of road freight transportation sector in the EU
De Feo *2019*	Efficiency of road freight	Journal	very high	Economic, energy and economic efficiency of road freight
Deineko *2023*	Analysis of the transportation system	Printed Book	very high	Exploratory analysis of the transport system for the modular vehicle concept Operated as passenger and freight transport using continuum approximation techniques.
Dellbügge *2023*	Resilience of urban factories and their supply chains	Printed Book	very high	Resilience of urban factories and their supply chains

Denteh *2025*	Integrating travel behavior forecasting	Journal	very high	Integrating travel behavior forecasting and generative modeling for predicting future mobility and spartial transportation
Deutsche Bahn *2024a*	Current DB Group reports	Home Page	high	The hydrogen train is expected to save 330 tons of CO_2 per year
Deutsche Bahn *2024b*	Sustainability	Home Page	high	Hydrogen on the rails
Deutsche Bundesstiftung Umwelt *2024*	Grundlagen einer nachhaligen Entwicklung	Home Page	high	
Deutsche Verkehrszeitung *2024*	Logistics objectives	Home Page	high	In order to counteract the overload of transport capacities, data analyses are used to optimize transport routes
DGG *2024*	Digitization in quality management and quality assurance	Home Page	high	
Digital Experts *2024*	Digitalization megatrend	Home Page	high	Digitalization will still be a megatrend in 2025, important for society
Dincer *2024*	Economic development and the envionmental ecosystem	Printed Book	very high	The Role of Energy Policy in Economic Growth
DIW *2024*	Energy and climate policy	Home Page	high	Energy and climate policy: more speed for the transformation
Diwo *2024*	Trends and megatrends in digitalization	Home Page	high	The secret drivers of digital transformation
Domagala *2023*	Economics	Home Page	high	Energy and environmental efficiency of road freight transportation sector in the EU
Dorfmann *2022*	European transport policy	Printed Book	very high	Mobility and digitalization are just some of the issues addressed by European transport policy

Duin *2023*	Planning of Hyper-loop-based cargo tubes	Printed Book	very high	Questions about capacity and operating costs need to be identified when planning the infrastructure of the Cargo Loop
DVZ *2023*	Growth in the volume of goods	Home Page	high	The increase in the volume of goods is taking on dramatic proportions
Ebert & Wu *2023*	Dynamic container routing problem	Printed Book	very high	Optimizing the flow of goods while avoiding capacity changes is a challenge
EEA *2024*	European Envinronmental Agency	Government	highest	EEA betont, dass der kombinierte Verkehr ein wesentliches Instrumend zur Erreichung der Klimaziele ist
Eisenkopf *2023*	Mobility on the climate test bench	Journal	very high	Mobility on the climate test rig, rail transport on the test rig
EIZ Rostock *2022*	Europe direct	Home Page	high	With an area of around ten million square kilometers, Europe is the second smallest continent in the world and is bordered to the west and north by the Atlantic Ocean
Elbasha & Abdellatif *2025*	AI-based smart traffic management system	Journal		IoT and technique
Ember *2025*	Mobility with a future	Home Page	high	Early end for internal combustion engines
e-Mobile *2024*	Climate protection in freight transport	Home Page	high	Importance of climate protection in freight transport
Energate *2024*	Testing hydrogen trains	Home Page	Level V	Hydrogen trains are subsidized by the state
Energy News *2024*	Siemens sees hydrogen as a future market	Home Page	Level V	Siemens sees hydrogen as a future market
Enste & Potthoff *2021*	Behavieral economics and climate protextion	Home Page	high	Bottom regulation and green nudges for more sustainability
Ernest & Young *2024*	Progress monitor for the decarbonization of energy	Home Page	high	Current key figures on the status of the decarbonization of energy and forecasts for upcoming investments

Ernste *2021*	Behavioral economics and climate protection	Home Page	high	Better regulation and green nudges for more sustainability
ETF *2018*	Studies on social and economic significance	Study	high	Studies on the social and economic importance of combined transport for employees in the transport industry
Euramco *2024*	Green Funds	Home Page	high	Green energy is becoming increasingly attractive
Euronews *2025*	The European tube	Home Page	high	Inside the project to launch a continentwide rail system by2040
European Commission *2019*	The European Green Deal	Government	highest	According to the European Commission, climate change is one of the greatest challenges and offers an opportunity to develop new economic models. The European Green Deal is the basis for the energy transition.
European Commission *2022*	The European Green Deal	Government	highest	
European Commission *2024*	A step towards climate neutrality	Government	highest	The EU is getting closer to its climate targets of being CO_2 neutral by 2050
European Commission *2024*	European Güner Deal	Government	highest	EU agrees stricter legal provisions to accelerate the expansion of renewable energy
European Commission *2023*	EU calls for combined transport	Government	highest	According to the European Commission, climate change is one of the greatest challenges and offers an opportunity to develop new economic models
European Commission *2024*	New target for 2040	Government	highest	The transport sector has a major role to play in climate change
European Committee *2019*	Standardization initiatives	Government	highest	Standardization initiatives to make combined transport more efficient
European Council *2024*	Climate+ neutrality	Government	highest	The European Council has adopted the EU Climate

				Change Act and sets binding targets
European Environment Agency *2024*	Climate neutrality in Europe	Government	highest	Opportunities and challenges of climate neutrality
European Environmental Agency 2019	Combined transport as a key argument for achieving climate targets	Government	highest	Member states were also required to submit a draft update of their climate plans reflecting the revised EU climate and energy targets for 2040
European Parliament *2024*	Greenhouse gas emissions	Government	highest	the European Parliament adopted the EU Climate Law, which sets a binding target of reducing emissions by 55% by 2030 and achieving climate neutrality by 2050
European Parliament *2016*	Combined Transport Directive	Government	highest	The review of the Combined Transport Directive 92/106/EEC in 2023 emphasizes the need to improve the efficiency and effectiveness of the directive
European Parliament *2023*	Green deal, key to a climate neutral and sustainable EU	Government	highest	The level of knowledge of the stakeholders must be determined in order to be able to initiate measures
European Parliament *2023*	Swedish freight transport systems	Government	highest	The creation of jobs and new professional fields for energy-efficient buildings and processes are further advantages
European Union *2012*	Treaty on the Functioning of the EU	Government	Highest	The EU consists of 27 member states, population of half a billion people
EWE *2024*	What is green energy	Home Page	high	Green Energy stands for clean air
Federal Foundation for the Environment *2024*	Fundamentals of sustainable development	Home Page	high	
Federal Government *2024*	EU climate protection package	Government	Highest	Fit for 55

Federal Statistical Office *2021*	Freight transport in the EU	Government	highest	trends and the associated political decisions will contribute significantly to the fact that road transport will continue to grow, while the shift to more environmentally friendly modes of transport such as rail is progressing more slowly
Federal Statistical Office *2024a*	Freight transport volumes	Government	highest	Greenhouse gases are mainly released through the use of fossil fuels
Federal Statistical Office *2024b*	Freight transport by mode of transport	Government	highest	Freight transport comprises 1.2 million companies and around 10.5 million employees
Federal Statistical Office *2024c*	European Green Deal	Government	highest	States commit to reducing global warming to below 2°C
Federal Statistical Office *2024d*	Volume and performance of traffic	Government	highest	Transport performance in Germany in figures
Federal Statistical Office *2022*	Freight performance of rail and inland waterway vessels	Government	highest	The transport performance of inland navigation has been falling continuously since 2011
Federal Statistical Office *2025*	*Beförderungsmenge und Beförderungsleistung*	Home Page e	high	*nach Verkehrsträgern*
Fedoskin et al. *2024*	Formation of a system	Home Page	high	Formation of a system of indicators for assessing the availability and results of using freight transport
Financial Times *2024*	EU pushes for tech transfers from Chinese clean tech firms	Home Page	High	
First Climate *2024*	Renewable power for companies	Home Page	high	Since no single sustainable energy source can meet all energy needs, a combination of different green, clean and renewable energy sources is necessary
Fischer *2021*	Climate protection strategies	Home Page	high	Climate protection strategies in the transport sector

Fischer *2024*	Innovations for green logistics	Printed Book	very high	Achieving a reduction in emissions with innovative transportation solutions
Fit for 55 *2022*	The EU's plan for a smooth transition	Home Page	high	Between 2010 and 2019, emissions from heavy goods vehicles fell by 31.1 percent
Fleischmann & Klose *2021*	Quantitative Approaches in Logistics	Printed Book	very high	Cargo tubes offer the possibility to reduce transport costs, energy losses can be minimized
Fockenbrock *2020*	Testing the hydrogen train	Home Page	high	DB deploys hydrogen-powered test trains together with Siemens
Fraunhofer *2024*	Mobility concepts for companies	Home Page	high	Mobility concept based on the individual needs of companies
Fraunhofer IFAM *2024*	Climate+ friendly hydrogen	Home Page	high	The focus is on the further development of electric vehicles and other environmentally friendly drive technologies such as hydrogen and fuel cell technologies
Fraunhofer ISE *2024a*	Paths to a climate-neutral energy system	Home Page	high	The Fraunhofer Institute develops measures for long-term corporate success. The mobility concept is based on the individual needs of companies
Fraunhofer ISE *2024b*	55% by 2023 and 100% by 2050	Home Page	high	In the previous elaboration of this work, it became clear that freight transport in Europe will increase sharply in the coming years, so a massive increase in CO2 emissions and other nitrogen oxides of up to 45% is also expected. At the same time, climate targets are to be achieved to reduce CO2 emissions by 55%
GAG *2024*	On the way to a sustainable global economy	Home Page	high	
Gatta et al. *2019*	Sustainable urban freight transport	Home Page	high	Sustainable urban freight transport adopting public transport-based

				crowdshipping for B2C deliveries
Gattorna *2019*	Digital transformation of supply chain management	Printed Book	very high	Digital transformation of supply chain management
Gavanas *2024*	New mobility services and sustainable urban development	Home Page	high	The common EU vision and the challenges for transport planning in Greek cities
German Association for Foreign Business *2024*	On the way to a more sustainable global economy	Home Page	high	On the way to a more sustainable global economy
German Federal Environmental Foundation *2024*	Fundamentals of sustainable development	Home Page	high	Fundamentals of sustainable development
German Institute for Economic Research *2024*	Energy and climate policy	Home Page	high	More speed for the transformation
German Society for Quality *2024*	Digitization in quality management	Home Page	high	Digitization in quality management
Gilbert *2018*	Transport Revolutions	Printed Book	very high	Moving People and Freight Without Oil
GIZ *2017*	Sustainable transportation	Printed Book	very high	A Sourcebook for Policy-Makers in Developing Cities
Gomes *2023*	Strategic organizational sustainability in the age of sustainable development	Journal	very high	Strategic organizational sustainability in the age of Sustainable Development Goals. Sustainability
Gomez-Baggethun *2023*	Economical Economics	Printed Book	very high	emphasize that ecological economics challenges the conventional economic paradigm by placing ecological limits at the forefront of economic analysis

Gonzalez-Feliu *2018*	Sustainable urban logistics	Printed Book	very high	Sustainable Urban Logistics, Concepts, Methods, and Information Systems
Göpfert *2024*	Logistics - Dynamic and innovative	Printed Book	very high	Freight transport is a part of transportation logistics that has a new management concept for the organization of goods and financial flows
Grant *2023*	European Sustainability	Printed Book	very high	Environmentally friendly SCM strategies contribute to competitiveness
Grant *2024*	Green supply chain management	Printed Book	very high	Reducing emissions is a key objective of sustainable mobility
Grant *2017*	The handbook of sustainable Logistics	Printed Book	very high	*The Handbook of Sustainable Logistics and Supply Chain Management*
Green Deal *2022*	Green deal	Home Page	high	
Greenpeace *2024*	Decarbonization of transport for climate and justice	Home Page	high	Additional overhead lines of 600 km per year are necessary
Grubb & Neuhoff *2023*	Economic pathways to carbon neutrality	Printed Book	very high	Market dynamics are not enough to achieve carbon neutrality
Güterbahnen *2023*	Zahl des Tages	Home Page	high	
H2-News *2024*	Hydrogen by rail	Home Page	high	Hydrogen by rail
Hall *2023*	Advances in green logistics	Printed Book	very high	The introduction of e-trucks is driving sustainable mobility forward
Hall *2022*	Handbook of transportation science	Printed Book	very high	Sustainable logistics solutions require extensive research and development efforts
Harris & Williams *2021*	Freight transportation	Journal	very high	Freight transportation, policy measures and market responses
Harttmann *2022*	Fuel cell truck	Home Page	high	fuel cell trucks will only be able to operate competitively in the medium term beyond niche applications

				with continued government intervention
Haug *2023*	2.2 degrees warmer	Home Page	high	Europe is warming twice as fast as other continents.
Henderson & Clark *1990*	Architectural innovation	Home Page	high	It is more difficult to implement change when resources are outdated and actors need to acquire new resources
Hensher & Button *2023*	Handbook of logistics	Printed Book	very high	Handbook of Logistics and Supply Chain Management
Hensher & Mulley *2022*	Public transportation	Printed Book	very high	improvement in air quality
Hettich *2023*	OEM and shipping behavior	Printed Book	very high	Delivery and shipping behavior of commercial freight Demand for logistics services Provider
Hinrichs *2021*	Sustainability as a corporate strategy	Printed Book	very high	Roadmap for corporate sustainability & innovation
Hoffmann *2021*	Sustainable mobility	Home Page	high	Sustainable mobility in climate protection in German cities
Hünding *2023*	Sustainability in manufacturing companies through digitalization	Printed Book	very high	Conceptualizing the use of digital technologies to achieve and improve sustainability goals
Hupac *2021*	Reports from a leading intermodal transport company on market conditions and future developments in combined transport.	Home Page	Level V	Reports from a leading intermodal transport company on market conditions and future developments in combined transport
Hydrogen Europe *2025*	Hydrogen refuling	Home Page	high	Hydrogen refuling projects allocation latest EU funding round
hySolutions *2024*	Hydrogen on the rails	Home Page	high	There are currently 15,000 diesel locomotives in operation
Ickert *2024*	Updating the perspectives on crossing the Alps	Printed Book	very high	Updating the prospects for transalpine freight transport

IEA *2025*	Renewabl	Home Page	High	The implementation of global value chains is not possible without the use of modern information and communication technologies
Inal *2024*	Decarbonization of shipping	Journal	very high	Hydrogen and fuel cells legislation in the maritime industry
Intermodal-news *2021*	Analysis of the performance and development of intermodal transport	Home Page	high	Latest news and analysis on the performance and development of intermodal transport in Europe
International Transport Forum *2025*	Sustainable Transport systems initiative	Home Page	high	
International Transportation *2019*	Combined transport	Journal	very high	
Iskan *2021*	Logistics sector for the future	Printed Book	very high	Between digital industrialization and analogue challenges
Izotov *2024*	Combined shipments container	Home Page	Level V	Multi–criteria assessment of the combined shipments container transportation possibility
Jackson *2023a*	Sustainable Freight Transport	Printed Book	very high	Sustainability can be achieved by reducing GHGs
Jackson *2023b*	Post Growth	Printed Book	very high	Sustainable growth on a finite planet requires a paradigm shift
Jakob *2023*	Paths to socially just decarbonization of transport	Printed Book	very high	Ways to a socially just decarbonization of transport: How we can make mobility sustainable
Janzen & Erdmann *2024*	Where there have been particularly severe fires in Europe	Home Page	high	Where there have been particularly severe fires in Europe
Jöhrens *2022*	Comparative analysis of the potential of drives	Home Page	high	The availability of cheap imported hydrogen must be considered questionable
Jones *2020*	Hydrogen vehicles	Printed Book	very high	Hydrogen vehicles in urban logistics: A total cost of ownership analysis and some policy implications

JUSDA *2024*	Emerging Trends in Land transport for 2025	Home Page	high	
Kaltenbrunn *2021*	Impacts of the Ecological Transformation on German Climate Protection	Home Page	high	The role of renewable energies in German climate protection
Kampp *2022*	Challenges in international long distance rail transport	Home Page	high	Challenges in international long distance rail transport in Europe
Kasten *2022*	Climate protection	Printed Book	very high	Climate protection in the transport sector
Kaszuba et al. *2024*	Sustainable mobility challenges	Home Page	high	Case study of the offshore center in Gdansk transport accessibility. *Economics and Environment*
Keller *2020*	Climate change in Germany	Journal	very high	Climate change strategies and water management in Germany
Kerry Health and Nutrition Institute *2025*	Ten key health and nutrition trends of this year	Home Page	high	
Kille *2024*	Freight transport compact	Printed Book	very high	In addition to a dense road network, Switzerland has one of the densest rail networks in the world
Kluth *2022*	Moving medium-term forecast for freight transport	Home Page	high	Moving medium-term forecast for freight transport
KOFA *2023*	Ecological sustainability	Home Page	high	Ecological change means looking at the challenges and opportunities in equal measure
Koneberg *2022*	Energy from wind and sun	Study	high	Energy from wind and sun, options for securing skilled workers
König *2021*	Climate protection in mediumsized German cities	Home Page	high	Climate protection in mediumsized German cities
Krauth *2025*	Sustainable freight trans-port systems and policies	Journal	very high	Optimising mode choice in a bi-modal freight network considering sustainability and urban logistic stakeholder perspective
Krüger *2017*	Municipal climate protection concepts	Journal	very high	Municipal climate protection concepts

Kummer *2023*	Low Carbon Logistics	Printed Book	very high	Logistics networks must be optimized to reduce carbon emissions
Laimer & Perathoner *2024*	Mobility and transportation law in Europe	Printed Book	very high	the expansion of sustainable transport networks is a necessary prerequisite for the functioning of the European single market
Lang *2021*	Inovative technologies for sustainable freigt transport	Journal	very high	Innovative technologies for sustainable freight transport
Letta et al. *2024*	Future of a single European market	Journal	very high	
Lexikon der Nachhaltigkeit *2024*	Sustainability	Home Page	Level V	Models and concepts for achieving sustainability
Liebisch-Gümüs *2024*	Mobilities and the history of mobility	Home Page	high	
Liu *2021*	A bi-level model for green freight transportation	Home-page	Level V	A bi-level model for green freight transportation pricing strategy
Lübben *2022*	Annual review, on the situation of the German aviation industry	Home Page	Level V	The state of the German aviation industry
Macharis *2021*	Sustainable Logistics and Transportation	Home Page	Level V	The environmental friendliness of cargo tubes is demonstrated by the significant reduction in air pollution
Mainzer *2021*	Artificial intelligence and sustainability	Printed Book	very high	Sustainability is a term with positive connotations
Manderscheid *2024*	Sustainable mobility and competition	Printed Book	very high	noise pollution for local residents and the continuing high number of deaths and injuries caused by traffic accidents in Europe
Manheller *2024*	VDV Annual Report	Home Page	Level V	VDV Annual Report
Mariana *2022*	Dimensions of freight transportation	Journal	very high	Economic, environmental, and social dimensions of freight transportation
Mayer *2019*	Climate protection measures	Home Page	high	Climate protection measures

McKinnon *2023*	The future of sustainable transport	Printed Book	very high	Integration of low GHG also reduces costs
McKinnon *2018*	The future of sustainable road freight	Journal	high	The use of innovative transportation solutions can reduce environmental pollution and the associated traffic congestion
Mehmedi *2021*	Electric vehicles	Printed Book	very high	Road transport is the most important mode of transport
Meissner *2024*	Ecological logistics	Printed Book	very high	Ecological logistics includes the reduction of emissions and the efficient use of resources
Meißner *2024*	Ökologische Logistik, Nachhaltigkeit im Güterverkehr	Printed book	very high	
Merz *2023*	Hidden champions	Printed Book	very high	Hidden champions in logistics
Monios *2023*	International freight transportation	Home Page	high	The environmentally friendly technology of cargo tubes helps to reduce CO2 emissions
Monsef *2022*	Employees in the renewable energy sector	Home Page	high	Employees in the renewable energy sector: production and manufacturing occupations in the focus of the decarbonization of energy
Muchna *2021*	Logistics basics	Home Page	very high	Transport services should be differentiated according to the individual modes of transport
Müller *2024*	Green hydrogen	Journal	very high	Powering the future of passenger and freight transport
Müller *2020*	Climate protection and the decarbonization of energy	Home Page	high	Climate protection and the decarbonization of energy in Germany. University of Hamburg
Nagabandi *2023*	3D printing	Home Page	high	How 3D printing can dramatically reduce carbon emissions in the manufacturing industry
Nature Sustainability	Progress towards the Sustainable	Home Page	High	Progress towards the Sustainable Development

2024	Development Goals.			Goals.
Navoco *2024*	Sustainability	Home Page	high	Sustainability
Nell *2024*	Mobility and logistics	Home Page	high	sustainable transportation concepts are needed that meet the requirements and environmental goals that have been set
Nickel *2023*	Mobility concept	Home Page	high	Success factors for the decarbonization of mobility are derived on the basis of mobility concepts
Novocargo *2024*	Last mile in urban areas the challenges	Home Page	high	
Now *2024*	Elektra II - Hydrogen barge	Home Page	high	Elektra II - Hydrogen barge
OECD *2025*	*The future of global value chains*	Homepage	high	Organisation for Economic Co-operation and Development
Öko-Institut *2024*	Fact check on nuclear energy	Home Page	high	Despite its CO2 neutrality, nuclear energy is often viewed critically, as its construction, transportation and final storage of fuel rods have an impact on ecosystems
Ordonez-Ponce *2023*	Green Transportation	Journal	very high	Green transportation for sustainability: Review of current barriers
Osztovics *2021*	Arena Analyse	Home Page	high	Marktwirtschaft neu
Palmetto *2025*	Green energies	Home Page	high	Hydropower plants generate renewable, but not clean, energy
Peng *2024*	Trends and economic challenges in green energy	Printed Book	very high	
Perevoznic *2024*	Archiving the 2030 agenda	Journal	very high	EU faces considerable challenges due to climate change
Perez *2022*	Reneawable Energy integration in freight	Journal	very high	

Perez & Rossi *2022*	Strategies for reducing CO2 emissions	Journal	very high	Strategies for reducing CO2 emissions in freight transportation
Pfaff *2020*	Praxishandbuch Marketing	Printed book	very high	ARENA Analyse
Piontek *2021*	Building blocks of logistics management	Printed Book	very high	Supply chain management, e-logistics, logistics controlling, green logistics, logistics instruments
Port of Rotterdam *2018*	Studies on the logistical advantages of combined transport	Study	high	Studies on the logistical advantages and infrastructure of combined transport in one of Europe's largest ports
Quaschning *2018*	Regenerative Energiesysteme	Printed book	very high	Technologie Bereichnung und Simulation
Quashning *2018*	Regenerative energy systems	Printed Book	very high	The amount of energy is measured in joules
Quentin *2023*	Transport planning beyond infrastructure change	Journal	very high	Transport planning is beyond infrastructural change. An empirical analysis of transport planning practices.
Racca et al. *2024*	Mapping past and future shipping noise in European seas.	Journal	very high	*The Journal of the Acoustical Society of America*
Rail system *2024*	Long distance rail transport in Europe	Home Page	high	Hundreds of tons of CO_2 are produced during the manufacture of steel for the rails
RailFreight.com *2024*	One billion Euro for smaller rail infrastructure	Home Page	high	
Rail-pro Alliance *2023*	Public service infrastructure	Home Page	high	Confederation needs clear goals
Rail-pro Alliance *2024*	Crossborder freight transport	Home Page	high	Electrification of border crossings must be driven forward
Railway Technology *2020*	Austria ÖBB launches hydrogen traint	Home Page	high	

Ramaekers *2018*	Using an integrated order picking-vehicle routing problem	Journal	very high	Using an integrated order picking-vehicle routing problem to study the impact of delivery time windows in e-commerce
Rammert *2024*	Mobility instead of traffic	Printed Book	very high	Rammert points out the influencing dimensions of mobility, as they must be understood for fully-fledged mobility planning.
Rat der Europäischen Union *2024*	Strategies for achieving climate neutrality	Homepage	high	Procedure and strategies for achieving climate neutrality
Raza *2025*	The potential of Hyperloop in Future supply chains	Home Page	high	
Reissova et al. *2024*	Job attractiveness	Journal	very high	Job attractiveness and distance as predictors of willingness to move
REN21 *2024*	Global Futures Report	Home Page	High	Renewables for Sustainable Transport
Research and Markets *2024*	Autonomous last mile. Delivery Business alanlysis report 2024.	Home Page	high	Market to reach 6,2 billion @ by 2030
Reset digital *2024*	Renewable energies	Home Page	high	Use of renewable sources, sustainable products and climate-friendly employer services
Reuters *2025*	South Koreas deadly fixes made twice as likely by climate change researchers	Home Page	high	
Richter *2020*	Climate protection in the German construction industry	Home Page	high	Climate protection in the German construction industry
Risius *2023*	Ecological sustainability	Study	high	In addition to social, economic and ecological aspects, there are other aspects of sustainability
Rodrigue *2021*	The geography of transport systems	Printed Book	very high	How logistics systems work together

Rodrigue *2022*	Handbook of logistics	Printed Book	very high	Logistics Basics, Workflow and Business models
Ryley *2023*	Transportation and climate change	Printed Book	very high	Adapting to climate change is the biggest challenge for freight transport
S&P Global *2025*	Avoid greenwashing accusations	Home Page	high	Reputational damage can occur
Saake & Sommer *2023*	Forecasting method for capacity utilization	Printed Book	very high	Focus on individual transportation creates challenges
Sampson *2023*	Greening the supply chain	Printed Book	very high	Reduce operating costs and increase efficiency
Savy *2023*	Freight transportation	Printed Book	very high	Cargo Tubes offer an advanced solution for modern freight transport
Schiffer *2023*	Introduction to the energy industry	Printed Book	very high	Introduction to the energy industry: resources and markets
Schmidt *2019*	Energy efficiency as the key to climate protection	Journal	very high	Green logistics: approaches to improving the environmental balance in freight transport
Schmitz *2023*	Decarbonization of freight transport	Printed Book	very high	Sustainable transport solutions and innovative drives are key elements of climate-friendly freight transport
Schneider *2024*	Green transport chains	Printed Book	very high	Increasing efficiency helps to reduce environmental impact
Schultze *2023*	Sustainable transport logistics	Printed Book	very high	Concepts and strategies for sustainable transport logistics
ScienceDirect *2020*	Importance of combined transport in Europe	Homepage	high	Importance and development of combined transport in Europe, including growth trends and infrastructure improvements
ScienceDirect *2021*	Rail transportation	Homepage	high	Siemens Mobility is testing the use of hydrogen for rail transportation. The combi-

				nation of a newly developed train and a mobile filling station is being tested.
Seeburger *2024*	Green logisctics - How digitalization is driving sustainability forward	Homepage	high	The concept of logistics is expanded to include the ecological aspect by integrating sustainable practices
Siebenfeiffer *2021*	Mobility of the future	Printed Book	very high	Urbanization increases from 55% to 68
Siebenpfeiffer *2021*	Mobilität der Zukunft	Printed book	very high	Intermodales Verkehrskonziept
Standage *2020*	The future of transportation	Printed Book	very high	Cooperation between technology and politics requires
Steininger *2022*	National energy and climate plan	Printed Book	very high	Many efforts and immediate action are needed to counteract rising GHGs
Stelling & Brunner *2023*	Exploring knowledge gaps	Printed Book	very high	Exploring knowledge gaps between key stakeholders in the transition to an electrified freight transport system in Sweden
Stevens *2023*	Advancements in electric and hydrogen fuel cell vehicles	Journal	very high	Renewable and Sustainable Energy Reviews
Stölzle *2024*	Freight transport compact	Printed Book	high	More than 720,000 vehicles offer loading capacity for more than 10.3 million tons of goods
System Bahn *2024*	Schienenverkehr in Europa	Home Page	high	Herausforderungen und Perspektiven
Szaruga et al. *2024*	The role of EU transport market liberalization	Home Page	high	The role of EU transport market liberalization in shaping directions of rail energy consumption rationalization in relation to the export of goods The case of Poland
Tagesschau *2024a*	Hydrogen for the railways	Home Page	high	hydrogen is still in very short supply for the foreseeable future and is urgently needed elsewhere,

				for example in industry or to power large ships
Tagesschau *2024b*	How the expansion of renewable energy is going	Home Page	high	"Waste" also includes materials such as waste wood or straw, which could be put to better use elsewhere
Texas A&M Transportation Institute *2023*	Transport Institute	Home Page	high	
Thakur *2025*	Rising network bandwidth	Home Page		Rising network bandwidth, trends and best practices
The freight railroads *2023*	Overview of the number of the day	Home Page	high	Overview of the number of the day
The Sustainable Agency *2025*	Greenwashing examples for 2024 and 2025	Home Page	high	
Tovias *2024*	Will the EU's fourth freedom be further challenged	Journal	high	Will the EU's fourth freedom be further challenged by present members of the single market
Transport club 2023	Association for nature and environmental protection	Home Page	high	Over 1.3 million trucks drive on the highways every day, with around a third coming from European countries
Transportation & Environment *2020*	Advocacy reports on the promotion of combined transport	Home Page	high	Advocacy reports to promote combined transport as a means of reducing traffic congestion and pollution
Tumlin *2024*	Sustainable Transportation Planning	Printed Book	very high	Joint planning by all stakeholders is crucial for the implementation of new strategies
UBA Dessau *2023*	Reduce GHG by 1.9%	Home Page	high	Stricter emission targets need to be introduced, countries' starting positions need to be adjusted
UIC *2021*	Role of combined transport	Home Page	high	Reports on the role of combined transport in the European Green Deal and

				the reduction of CO2 emissions
UIRR *2020*	Report on combined transport	Home Page	high	International Road Transport Union for Road-Rail Combined Transport. 2020 report on combined transport in Europe
Umbach *2023*	Energy security under the conditions of decarbonization	Journal	very high	Energy security under conditions of decarbonization of the economy and transport
Umweltbundesamt *2024*	Digital sustainability	Printed Book	very high	Electricity generation, heat generation and transport activities have a major impact on the environment due to greenhouse gas emissions
Umweltbundesamt *2023*	Research program 2023-2027	Home Page	highest	CO_2 emissions continue to rise
Umweltbundesamt *2022*	The European Green Deal	Home Page	highest	Means of transport can be better compared in terms of use and relevance using formulas and indices
Umweltbundesamt *2024*	Digital sustainability, sustainable design	Home Page	highest	Freight transport is the third largest producer of GHGs
Umweltbundesamt *2024*	Digital mobility platform	Home Page	highest	Generating electricity and heat pollutes the environment
Umweltbundesamt *2024*	Goals of the Paris Climate Agreement	Home Page	highest	It is becoming increasingly difficult to achieve environmental targets
Umweltbundesamt *2024*	Climate protection in transport	Home Page	highest	The increase in freight transport is countered by climate targets, which aim to reduce GHG emissions by up to 55% in the same period
Umweltbundesamt *2024*	Environmentally friendly freight transport	Home Page	highest	Freight transport has become significantly more efficient since the early 1990s, but the increase in transport performance has led to an increase in final energy consumption
UN environment *2021*	Green energy choices	Home Page	highest	Green Energy differentiates itself from clean and renewable energy

United Nations Department of Economic and Social Affairs *2025*	World social report	Govern-ment	highest	A new consensus to accelerate social progress
Unsere Welt in Daten *2023*	Greenhouse gas emissions in figures	Home Page	High	Greenhouse gas emissions by sector
Unterbroek *2023*	Rail transport across the Alps stagnates	Home Page	high	Rail transport across the Alps stagnates
Varnavskii *2024*	EU railways	Journal	very high	Contribution to decarbonisation solution. *Contemporary Europe*
VCD *2022*	Freight transport endangers the climate	Home Page	high	Companies are concerned that they appear more environmentally friendly to the outside world than they are
VDE *2022*	Hydrogen	Home Page	high	Study on hydrogen resources in rail freight transport
VDI *2024*	Logistics initiative	Home Page	high	Logistics initiative aims to get more freight onto rail and inland waterways
VDV *2024*	Giving priority to rail transport	Home Page	high	Demand for savings and management systems
VKU *2024*	Effects of the EU climate target	Home Page	high	Effects of the EU climate target
Von Malmborg *2024*	At the controls	Journal	high	Politics and policy entrepreneurs in EU policy to decarbonize maritime transport
Waalany *2021*	PESTLE Analyssis	Home Page	high	
Wagner *2023*	Sustainable freight transport concepts	Printed Book	very high	transport volumes are expected to continue to rise in the coming years
Wagner *2024*	Mobility through the ages	Printed Book	very high	Mobility concepts will not become self-sufficient
Weber *2019*	Renewable energy and grid stability	Home Page	high	Renewable energy and grid stability in Germany

Weber *2023*	Sustainability management in logistics	Printed Book	very high	One trend is the development of sustainable logistics concepts through innovative products such as ZS
Weidemann 2022	Emission-free and congestion-free transalpine freight traffic	Printed Book	very high	The slow expansion of the roads is offset by a large increase in the transportation capacity of heavy goods vehicles
Wensauer *2023*	Sustainable logistics	Printed Book	very high	Strategies for an ecological future
Wieland *2021*	Two perspectives on supply chain resilience	Journal	very high	Two perspectives on supply chain resilience
Wilde *2023*	New mobility	Printed Book	very high	Factors, drivers and changes
Wired *2023*	Sustainability management	Printed Book	very high	It is conceivable that politics and society will expand the requirements
Witte *2024*	Why rail hardly competes with trucks	Home Page	high	Rail freight transport is mainly profitable for long distances
World Economic Forum *2025*	The Henderson-Clark Innovation model	Home Page	high	The Henderson-Clark Innovation model
World of Work *2024*	The Henderson-Clark innovation model	Home Page	high	Actors often have to acquire new resources
Wozniak-Jechorek *2024*	Examining the European transition	Journal	very high	Central Eastern European countries have two decades of post-EU accession
Wright *2023*	Sustainable Freight Transport, Theory and modeling	Printed Book	very high	Modeling helps to quantify the impact of sustainable transport strategies
Wu et al. *2024*	Mandatory disclosure and corporate ESG performance	Journal	very high	Evidence from China's *explanation for nondisclosure* requirement.
Xiong et al. *2024*	Calculation of mixed oil volume	Journal	very high	Calculation of mixed oil volume for sequential transportation of crude oil pipelines with variable temperature and low Reynolds number

Yunex *2025*	ITS Industry in the Intelligent Age	Home Page	high	Trends of ITS Industry in the Intelligent Age
Yunex Traffic *2025*	Industry in the intelligent age trend report	Home Page	high	Industry in the intelligent age trend report
Zielinski *2024*	Green Transportation	Printed Book	very high	Technologies, Policies and Practices

Table 23: Model research table
Source: Own illustration.

The synthesis matrix shows, in which the author organized and analyzed the research by indicating whether a given finding or result occurred in a particular study or article.

Government = highest
Journals, books and articles = very high
company pages = high
other information = medium

Appendix 14: Model synthesis table

Source	Purpose	Mobility	Ecologigal	Social	Economical	Technical	Green energy
Accenture *2024*	Climate effects of digitalization			x		x	
Adams *2019*	Green freight transportation		x	x	x		x
Alekseev *2024*	Assessment of the modern climate policy						x
Bagwan *2023*	Refrigeration technologies in refrigerated supply chains					x	x
Baloise *2024*	What do we mean by mobility	x				x	
Banister *2022*	Unsustainable transportation	x	x	x	x	x	x
Banister *2023*	Sustainable transportation	x	x	x	x	x	x
BDE *2023*	European Commission				x		x
Becker *2019*	Impacts of the Ecological Transformation on Climate Protection		x				x
Behrends *2022*	Transportation and Lotistics Planning	x					
Black *2023*	The green logistics handbook	x					x
Blauwens *2022*	Transportation	x			x		
BMDV *2023*	The future needs paths	x					x
BMWK *2024a*	Reneawable Energy						x
BMWK *2024b*	Hydrogen for waterways	x				x	
Bothe *2024*	Energy Transition in Everyday Understanding	x					x
Bouchery *2017*	Sustainable supply chain	x					
Bräkling *2020*	Logistics management	x				x	
Brecha *2023*	Decarbonizing road freight transport	x					
Brown *2021*	The role of energy in German climate protection		x	x	x	x	
Browne *2023a*	Decarbonizing	x					
Browne *2023b*	Urban freight	x	x		x		
Brunner *2020*	Impact of emerging transport	x					
BSL *2020*	Combined transport	x			x		

Bundesklimaschutzgesetz *2019*	Climate Protection Act	x	x				x
Business.gov *2025*	Zero-emssions				x		x
BVL *2024*	Logistics objectives	x					
Caldeirinha et al. *2023*	Role of cargo owner in Logistic chain sustainability				x		x
Calderinha *2023*	Role of cargo owner in	x				x	
Carter *2023*	Climate Change and maritime transport				x		x
CISL *2025*	Sustainability in water unprecedated attack		x	x	x	x	X
Climate Portal *2024*	Reneawable Energy	x					x
CM Logistics *2024*	Freight transport: rail versus truck	x	x	x			
Cole *2023*	Eco-friendly transport logistics	x					x
Consultport *2024*	7S Modell nach McKinsey						
Courtier *2021*	Sustainable mobility and climate protection	x			x		
Courtier *2023*	Green Logistics	x					
Coyle *2023*	Transportation	x					
Crainic *2023*	Intermodal transportation	x					
Crawford *2024*	The green transition						x
CT4EU *2022*	Intelligent networking					x	
Cullinane *2024*	Innovations in green logistics	x					
Curtis *2024*	Handbook of sustainable Transport	x	x	x	x		x
Dabia & Nastase *2024*	Sylvicultura aeconomica	x					
Dabija & Nastase *2024*	Exploring the evolution of sustainability paradigms		x	x	x	x	
Davis et al. *2024*	Smart automated solution for public transport system	x	x				
De Feo *2019*	Economic transportation	x			x		
De Feo *2019*	Efficiency of road freight	x				x	
Deineko *2023*	Analysis of the transportation system	x				x	
Dellbügge *2023*	Resilience of urban factories and their supply chains	x					x
Delrada *2024*	Climate+ Technology, Clean Energy					x	x
Denteh *2025*	Integrating travel behavior forecasting	x	x				
Deutsche Bahn	Current DB Group reports	x	x	x	x		

2024a							
Deutsche Bahn *2024b*	Sustainability		x	x	x		
Deutsche Bundesstiftung Umwelt *2024*	Grundlagen einer nachhaligen Entwicklung		x	x	x	x	
DGG *2024*	Digitization in quality management and quality assurance					x	
Digital Experts *2024*	Digitalization megatrend					x	
Dincer *2024*	Economic development and the envionmental ecosystem				x		
DIW *2024*	Energy and climate policy						x
Diwo *2024*	Trends and megatrends in digitalization					x	
Domagala *2023*	Economic, energy and environmental efficiency of road freight transportation sector in the EU				x		
Dorfmann *2022*	European transport policy	x					
Duin *2023*	Planning of Hyperloop-based cargo tubes	x			x		
DVZ *2023*	Growth in the volume of goods	x					
Ebert & Wu *2023*	Dynamic container routing problem	x				x	
EEA *2024*	European Envinronmental Agency	x	x	x	x		
Eisenkopf *2023*	Mobility on the climate test bench	x					
EIZ Rostock *2022*	Europe direct	x					
Elbasha & Abdellatif *2025*	AIoT-based smart traffic management system		x			x	
Ember *2021*	Mobility with a future	x				x	x
e-Mobile *2024*	Climate protection in freight transport	x					x
Energate *2024*	Testing hydrogen trains					x	x
Energy News *2024*	Siemens sees hydrogen as a future market					x	x
Enste & Potthoff *2021*	Behavieral economics and climate protextion		x			x	x
Ernest & Young *2024*	Progress monitor for the decarbonization of energy	x					x
Ernste *2021*	Behavioral economics and climate protection			x			
ETF *2018*	Studies on social and economic significance		x		x		
Euramco *2024*	Green Energy - green funds						x
Euronews *2025*	The European tube					x	

Europ. Shippers Council *2019*	Strengthening more combined transport in Europe	x					
European Commission *2019*	The European Green Deal	x					x
European Commission *2022*	The European Green Deal	x					x
European Commission *2024*	A step towards climate neutrality	x					x
European Commission *2024*	European Green Deal	x					x
European Commission *2023*	EU calls for combined transport	x					
European Commission *2024*	New target for 2040	x			x		x
European Committee *2019*	Standardization initiatives	x					
European Council *2024*	Climate+ neutrality	x					x
European Environment Agency *2024*	Climate neutrality in Europe	x					x
European Environmental Agency *2019*	Combined transport as a key argument for achieving climate targets	x	x	x			
European Parliament *2024*	Greenhouse gas emissions		x		x		
European Parliament *2016*	Combined Transport Directive	x	x	x			
European Parliament *2023*	Green deal, key to a climate neutral and sustainable EU	x					
European Parliament 2023	Swedish freight transport systems	x					
European Union *2012*	Treaty on the Functioning of the EU	x					
EWE *2024*	What is green energy						x
Federal Government *2024*	EU climate protection package	x			x		x
Federal Statistical Office *2021*	Freight transport in the EU	x					
Federal Statistical Office *2024a*	Freight transport volumes	x					
Federal Statistical Office *2024b*	Freight transport by mode of transport	x					
Federal Statistical Office *2024c*	European Green Deal	x					x
Federal Statistical Office *2024d*	Volume and performance of traffic	x					
Federal Statistical Office *2022*	Freight performance of rail and inland waterway vessels	x					
Federal Statistical Office *2025*	*Beförderungsmenge und Beförderungsleistung*						
Fedoskin et al.	Formation of a system	x				x	x

2024							
Financial Times *2024*	EU pushes for tech transfers from Chinese clean tech firms			x			
First Climate *2024*	Renewable power for companies						x
Fischer *2024*	Innovations for green logistics						x
Fischer *2021*	Climate protection strategies	x					
Fit for 55 *2022*	The EU's plan for a smooth transition	x					
Fleischmann & Klose *2021*	Quantitative Approaches in Logistics	x					
Fockenbrock *2020*	Testing the hydrogen train	x				x	x
Fraunhofer *2024*	Mobility concepts for companies	x					x
Fraunhofer IFAM *2024*	Climate+ friendly hydrogen					x	x
Fraunhofer ISE *2024a*	Paths to a climate-neutral energy system	x			x		x
Fraunhofer ISE *2024b*	55% by 2023 and 100% by 2050	x			x		x
GAG *2024*	On the way to a sustainable global economy		x	x	x	x	
Gatta et al. *2019*	Sustainable urban freight transport	x	x	x	x	x	x
Gattorna *2019*	Digital transformation of supply chain management	x				x	
Gavanas *2024*	New mobility services and sustainable urban development	x		x			
German Association for Foreign Business *2024*	On the way to a more sustainable global economy		x	x	x	x	
German Federal Environmental Foundation *2024*	Fundamentals of sustainable development		x	x	x	x	
German Institute for Economic Research *2024*	Energy and climate policy						x
German Society for Quality *2024*	Digitization in quality management					x	x
Gilbert *2018*	Transport Revolutions	x					
GIZ *2017*	Sustainable transportation	x					
Gomes *2023*	Strategic organizational sustainability in the age of sustainable development	x			x	x	x
Gomez-Baggethun *2023*	Economical Economics				x		
Gonzalez-Feliu *2018*	Sustainable urban logistics	x					

Göpfert *2024*	Logistics - Dynamic and innovative	x				x	
Grant *2023*	European Sustainability	x	x	x	x	x	x
Grant *2024*	Green supply chain management						x
Grant *2017*	The handbook of sustainable Logistics	x					x
Green Deal 2022	Green deal	x					x
Greenpeace *2024*	Decarbonization of transport for climate and justice	x					x
Grubb & Neuhoff *2023*	Economic pathways to carbon neutrality				x		x
Güterbahnen *2023*	Zahl des Tages		x		x		
H2-News *2024*	Hydrogen by rail	x				x	x
Hall 2023	Advances in green logistics						x
Hall *2022*	Handbook of transportation science	x					
Harris & Williams *2021*	Freight transportation	x					
Harttmann *2022*	Fuel cell truck	x				x	x
Haug *2023*	2.2 degrees warmer						x
Henderson & Clark *1990*	Architectural innovation	x				x	
Hensher & Button *2023*	Handbook of logistics	x					
Hensher & Mulley *2022*	Public transportation	x					
Hettich *2023*	OEM and shipping behavior			x			
Hinrichs *2021*	Sustainability as a corporate strategy	x	x	x	x	x	
Hoffmann *2021*	Sustainable mobility	x	x	x	x	x	
Hünding *2023*	Sustainability in manufacturing companies through digitalization	x				x	
Hupac *2021*	Reports from a leading intermodal transport company in combined transport.	x					
Hydrogen Europe *2025*	Hydrogen refuling					x	x
hySolutions *2024*	Hydrogen on the rails	x				x	x
IEA *2025*	Renewables. International Energy Agency	x	x	x	x	x	
Inal *2024*	Decarbonization of shipping	x					x
Intermodalnews *2021*	Analysis of the performance and development of intermodal transport	x					

Internat. Transportation *2019*	Combined transport	x			x		
International Transport Forum *2025*	Sustainable Transport systems initiative		x	x	x	x	y
International weekly newspaper	Logistics objectives	x					
Iskan *2021*	Logistics sector for the future	x					x
Izotov *2024*	Combined shipments container	x					
Jackson *2023a*	Sustainable Freight Transport	x	x	x	x		
Jackson *2023b*	Post Growth	x				x	x
Jakob *2023*	Paths to socially just decarbonization of transport	x		x			
Janzen & Erdmann *2024*	Where there have been particularly severe fires in Europe		x				
Jöhrens *2022*	Comparative analysis of the potential of drives					x	x
Jones *2020*	Hydrogen vehicles	x				x	x
JUSDA *2024*	Emerging Trends in Land transport for 2025	x				x	x
Kampp *2022*	Challenges in international long distance rail transport	x					
Kasten *2022*	Climate protection in the transport sector	x					x
Kaszuba et al. *2024*	Sustainable mobility challenges	x	x	x	x	x	
Keller *2020*	Climate change in Germany	x					x
Kerry Health and Nutrition Institute *2025*	Ten key health and nutrition trends of this year		x	x	x		
Kille *2024*	Freight transport compact	x					
Kluth *2022*	Moving medium-term forecast for freight transport	x					
KOFA *2023*	Ecological sustainability		x				
Koneberg *2022*	Energy from wind and sun	x			x		
König *2021*	Climate protection in medium-sized German cities	x					x
Krauth *2025*	Sustainable freight transport systems and policies		x		x		
Krüger *2017*	Municipal climate protection concepts						x
Kummer *2023*	Low Carbon Logistics	x	x				x
Laimer & Perathoner *2024*	Mobility and transportation law in Europe	x					

Lang *2021*	Inovative technologies for sustainable freigt transport					x	
Letta et al. *2024*	Future of a single European market	x				x	x
Lexikon der Nachhaltigkeit *2024*	Sustainability		x	x	x	x	
Liebisch-Gümüs *2024*	Mobilities and the history of the mobility	x					
Liu *2021*	A bi-level model for green freight transportation	x					x
Lübben *2022*	Annual review, on the situation of the German aviation industry	x			x		
Macharis *2021*	Sustainable Logistics and Transportation	x	x	x	x		
Mainzer *2021*	Artificial intelligence and sustainability				x		
Manderscheid *2024*	Sustainable mobility and competition	x					x
Manheller *2024*	VDV Annual Report	x	x	x			
Mariana *2022*	Dimensions of freight transportation	x					
Mayer *2019*	Climate protection measures						x
McKinnon *2023*	The future of sustainable transport	x					x
McKinnon *2018*	The future of sustainable road freight	x			x		
Mehmedi *2021*	Electric vehicles	x				x	x
Meissner *2024*	Ecological logistics		x				
Meißner *2024*	Ökologische Logistik, Nachhaltigkeit im Güterverkehr				x		
Merz *2023*	Hidden champions	x					x
Meyer *2019*	Climate policy and effects						x
Monios *2023*	International freight transportation	x					
Monsef *2022*	Employees in the renewable energy sector			x			
Muchna *2021*	Logistics basics	x					
Müller *2024*	Green hydrogen					x	x
Müller *2020*	Climate protection and the decarbonization of energy	x					x
Nagabandi *2023*	3D printing					x	
Nature Sustainability *2024*	Progress towards the Sustainable Development Goals.			x			
Navoco	Sustainability		x	x	x	x	

2024							
Nell *2024*	Mobility and logistics						
Nickel *2023*	Mobility concept	x					x
Novocargo *2024*	Last mile in urban areas the challenges	x					x
Now *2024*	Elektra II - Hydrogen barge	x				x	x
OECD *2025*	The future of global value chains	x					
Öko-Institut 2024	Fact check on nuclear energy	x					
Ordonez-Ponce *2023*	Green Transportation	x					x
Osztovics *2021*	Arena Analyse						
Palmetto *2025*	Green energies					x	
Peng *2024*	Trends and economic challenges in green energy						
Perevoznic *2024*	Archiving the 2030 agenda	x			x		
Perez *2022*	Reneawable Energy integration in freight		x	x	x		x
Perez & Rossi *2022*	Strategies for reducing CO2 emissions		x			x	
Pfaff *2020*	Praxishandbuch Marketing						
Piontek *2021*	Building blocks of logistics management	x					
Port of Rotterdam *2018*	Studies on the logistical advantages of combined transport	x					
Quaschning *2018*	Regenerative Energiesysteme						
Quashning *2018*	Regenerative energy systems	x				x	x
Quentin *2023*	Transport planning beyond infrastructure change	x					
Racca et al. *2024*	Mapping past and future shipping noise in European seas	x					
Rail system *2024*	Long distance rail transport in Europe	x					
RailFreight.com *2024*	One billion Euro for smaller rail infrastrukture	x		x		x	
Rail-pro Alliance *2023*	Public service infrastructure			x		x	
Rail-pro Alliance *2024*	Cross-border freight transport	x	x				
Railway Technology *2020*	Austria ÖBB launches hydrogen train	x				x	x
Ramaekers *2018*	Using an integrated order picking-vehicle routing problem	x					
Rammert *2024*	Mobility instead of traffic	x					

Rat der europäischen Union *2024*	Strategies for achieving climate neutrality		x				x
Raza *2025*	The potential of Hyperloop in Future supply chains	x				x	xx
Reissova et al. *2024*	Job attractiveness			x			
REN21 *2024*	Status quo and prospects	x					
Research and Markets *2024*	Autonomous last mile. Delivery Business alanlysis report 2024.	x				x	
Reset digital *2024*	Renewable energies					x	x
Reuters *2025*	South Koreas deadly fixes made twice as likely by climate change researchers			x	x		
Richter *2020*	Climate protection in the German construction industry						x
Risius *2023*	Ecological sustainability		x				
Rodrigue *2021*	The geography of transport systems						
Rodrigue *2022*	Handbook of logistics	x					
Ryley *2023*	Transportation and climate change	x					x
S&P Global *2025*	Technology to reduce emissions and confront climate change		x				
Saake & Sommer *2023*	Forecasting method for capacity utilization	x					
Sampson *2023*	Greening the supply chain						x
Savy *2023*	Freight transportation	x					
Schiffer *2023*	Introduction to the energy industry		x				x
Schmidt *2019*	Energy efficiency as the key to climate protection						x
Schmitz *2023*	Decarbonization of freight transport	x					x
Schneider *2024*	Green transport chains	x					x
Schultze *2023*	Sustainable transport logistics	x					x
ScienceDirect *2020*	Importance of combined transport in Europe	x	x		x		
ScienceDirect *2021*	Rail transportation	x					x
Seeburger *2024*	Green logisctics - How digitalization is driving sustainability forward					x	x
Siebenfeiffer *2021*	Mobility of the future	x					
Siebenpfeiffer *2021*	Mobilität der Zukunft	x			x		
Standage *2020*	The future of transportation	x					

Steininger *2022*	National energy and climate plan	x					
Stelling & Brunner *2023*	Exploring knowledge gaps	x					
Stevens *2023*	Advancements in electric and hydrogen fuel cell vehicles					x	
Stölzle *2024*	Freight transport compact	x					
System Bahn 2024	Schienenverkehr in Europa		x		x		x
Szaruga et al. *2024*	The role of EU transport market liberalization						
Tagesschau *2024a*	Hydrogen for the railways	x					x
Tagesschau *2024b*	How the expansion of renewable energy is going					x	x
Texas A&M Transportation Institute *2023*	Transport Institute	x					
Thakur *2025*	Rising network bandwidth	x					x
than *2024*	Updating the perspectives on crossing the Alps	x					
The freight railroads	Overview of the number of the day in logistics	x					
The Sustainable Agency *2025*	Greenwashing examples for 2024 and 2025		x				
Tovias *2024*	Will the EU's fourth freedom be further challenged						
Transportation & Environment *2020*	Advocacy reports on the promotion of combined transport	x			x		
Tumlin *2024*	Sustainable Transportation Planning	x					x
UBA Dessau *2023*	Reduce GHG by 1.9%						x
UIC *2021*	Role of combined transport	x	x	x			
UIRR *2020*	Report on combined transport	x	x	x			
Umbach *2023*	Energy security under the conditions of decarbonization		x				x
Umeltbundesamt *2024*	Digital sustainability					x	
Umeltbundesamt *2023*	Reduction of GHG						x
Umeltbundesamt *2022*	The European Green Deal						x
Umeltbundesamt *2024*	Digital sustainability, sustainable design		x	x	x	x	
Umeltbundesamt *2024*	Digital mobility platform					x	
Umeltbundesamt *2024*	Goals of the Paris Climate Agreement						x
Umeltbundesamt *2024*	Climate protection in transport	x					x

Umeltbundesamt *2024*	Environmentally friendly freight transport	x					x
Umeltbundesamt *2024*	Fundamentals of sustainable development	x	x	x	x	x	x
UN environment *2021*	Green energy choices						x
United Nations Department of Economic and Social Affairs *2025*	World social report			x			
Unsere Welt in Daten *2023*	Greenhouse gas emissions in figures	x					x
Unterbroek *2023*	Rail transport across the Alps stagnates	x	x				
Varnavskii *2024*	EU railways	x					x
VCD *2022*	Freight transport endangers the climate	x	x				
VDE *2022*	VDI Study 2022	x					
VDI *2024*	Logistics initiative		x	x	x		
VDV 2024	Giving priority to rail transport	x					
Verkehrsclub *2023*	Association for nature and environmental protection						x
VKU *2024*	Effects of the EU climate target	x					x
Von Malmborg *2024*	At the controls						
Waalany *2021*	PESTLE Analyssis						
Wagner *2023*	Sustainable freight transport concepts	x					x
Wagner *2024*	Mobility through the ages			x		x	x
Weber *2019*	Renewable energy and grid stability	x				x	x
Weber *2023*	Sustainability management in logistics		x	x	x	x	
Weidemann *2022*	Emission-free and congestion-free transalpine freight traffic	x					x
Wensauer *2023*	Sustainable logistics	x					
Wieland *2021*	Two perspectives on supply chain resillience	x					x
Wilde *2023*	New mobility	x					
Wired *2023*	Sustainability management		x	x	x	x	
Witte *2024*	Why rail hardly competes with trucks	x				x	x
World Economic Forum *2025*	The Henderson-Clark Innovation model	x				x	x

World of Work *2024*	The Henderson-Clark innovation model					X	
Wozniak-Jechorek *2024*	Examining the European transition	x	X				
Wright *2023*	Sustainable Freight Transport, Theory and modeling	x	X			X	x
Wu et al. *2024*	Mandatory disclosure and corporate ESG performance			x			
Xiong et al. *2024*	Calculation of mixed oil volume	x					
Yunex *2025*	ITS Industry in the Intelligent Age					x	
Yunex Traffic *2025*	Industry in the intelligent age trend report	x				x	x
Zielinski *2024*	Green Transportation	x					x

Table 24: Model Synthesis Table
Source: Own illustration.

The research question determines which findings are relevant to the study. Use it as a guide to decide what should be included under each Findings header and how many findings are significant.

Appendix 15: Research poster (Draft)

1. Introduction

This research explores sustainable mobility in European freight transport using a qualitative approach to gain in-depth insights and derive optimization recommendations.

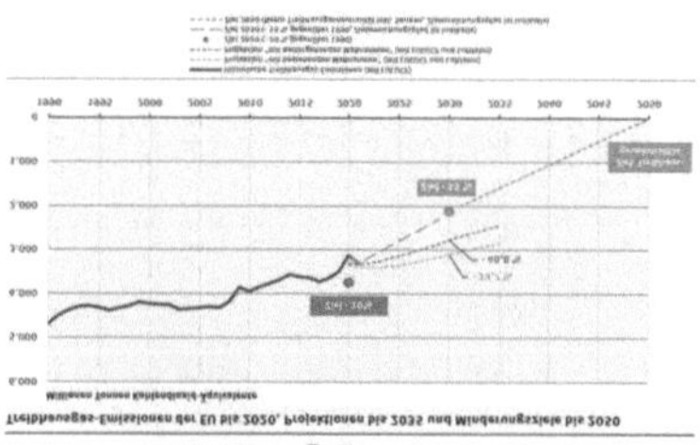

2 Research question

Key question

What are the potentials and risks of mobility competition?

Sub-questions

(1) To what extent and how must European freight transport adapt in order to achieve the sustainability targets by 2040?

(2) What specific measures can help to make European freight transport more sustainable in line with the climate targets that have been set?

(3) How can green energy be used sensibly in European freight transport?

The central question addresses potential and risks in mobility competition. Sub-questions focus on making freight transport more sustainable and the sensible use of green energy.

3. Methods

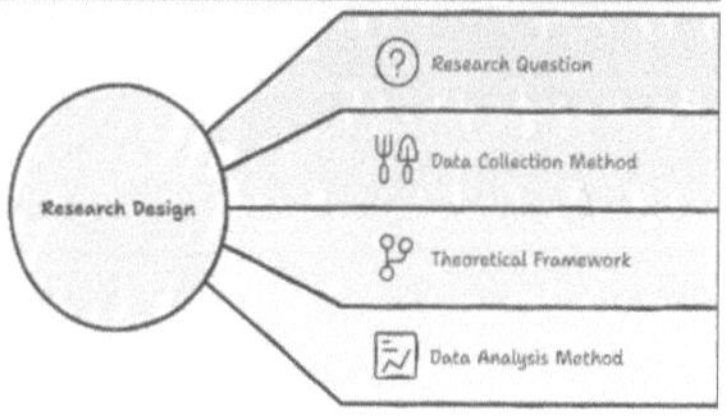

The qualitative approach enables a comprehensive analysis of complex relationships, focusing on identifying optimization potential and developing practical recommendations for sustainable freight transport.

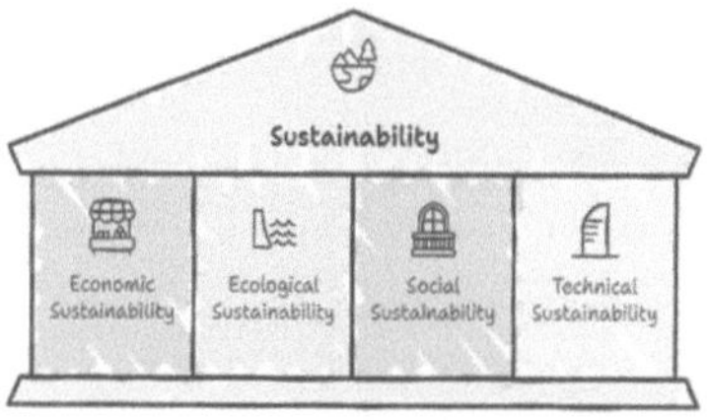

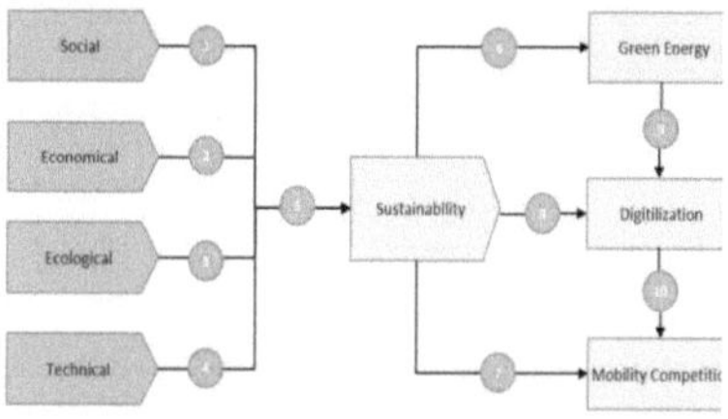

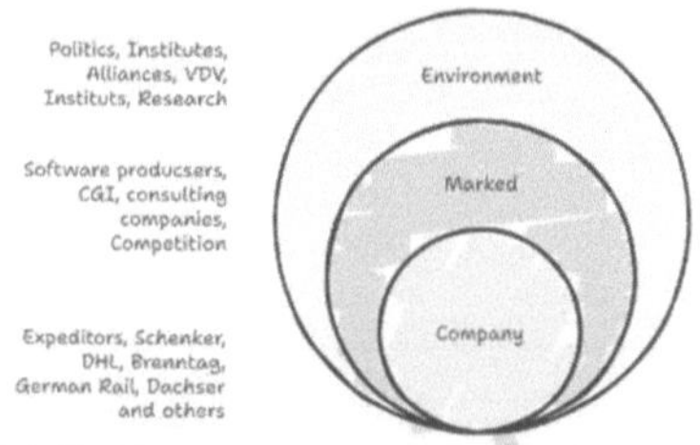

4. Literature Analysis

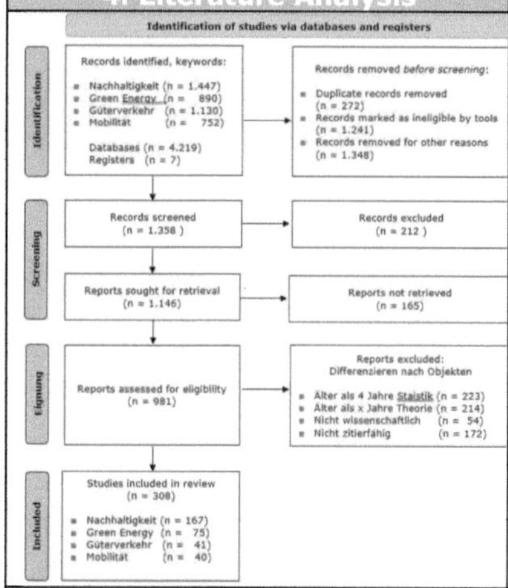

5. Results

The results show that sustainable freight transport in Europe is achievable through targeted measures, innovation, and stakeholder cooperation.

Future research should focus on implementing new technologies.

6. Conclusion

(1) Shift to rail
(2) Electrification & renewables
(3) Fleet modernization & alternative systems
(4) Green hydrogen & alliances
(5) Technological optimization
(6) Digitalization
(7) Policy & legal frameworks
(8) Sustainability labels & social justice
(9) Role of AI

7. Three Central keywords

(1) Interdisciplinary Decarbonization: Integrative model for CO_2 reduction in European freight transport.
(2) Sustainable Mobility Strategy: Concept for a resource-efficient, low-emission transport system.
(3) Tech-Policy Synergies: Interaction of innovation and policy to optimize freight transport.

8. Literature

- Traceability is ensured according to Gläser / Laudel 2010
- Semi-structured interviews with experts according to Misoch 2019
- Elaboration and transcription according to Dresing and Pehl 2018
- Deductive categories established on the basis of the theoretical analysis according to Mayring 2022
- Evaluation with qualitative content analysis according to Mayring 2022

Figure 22: Research Poster
Source: Own illustration.

Appendix 16: Research poster (English version)

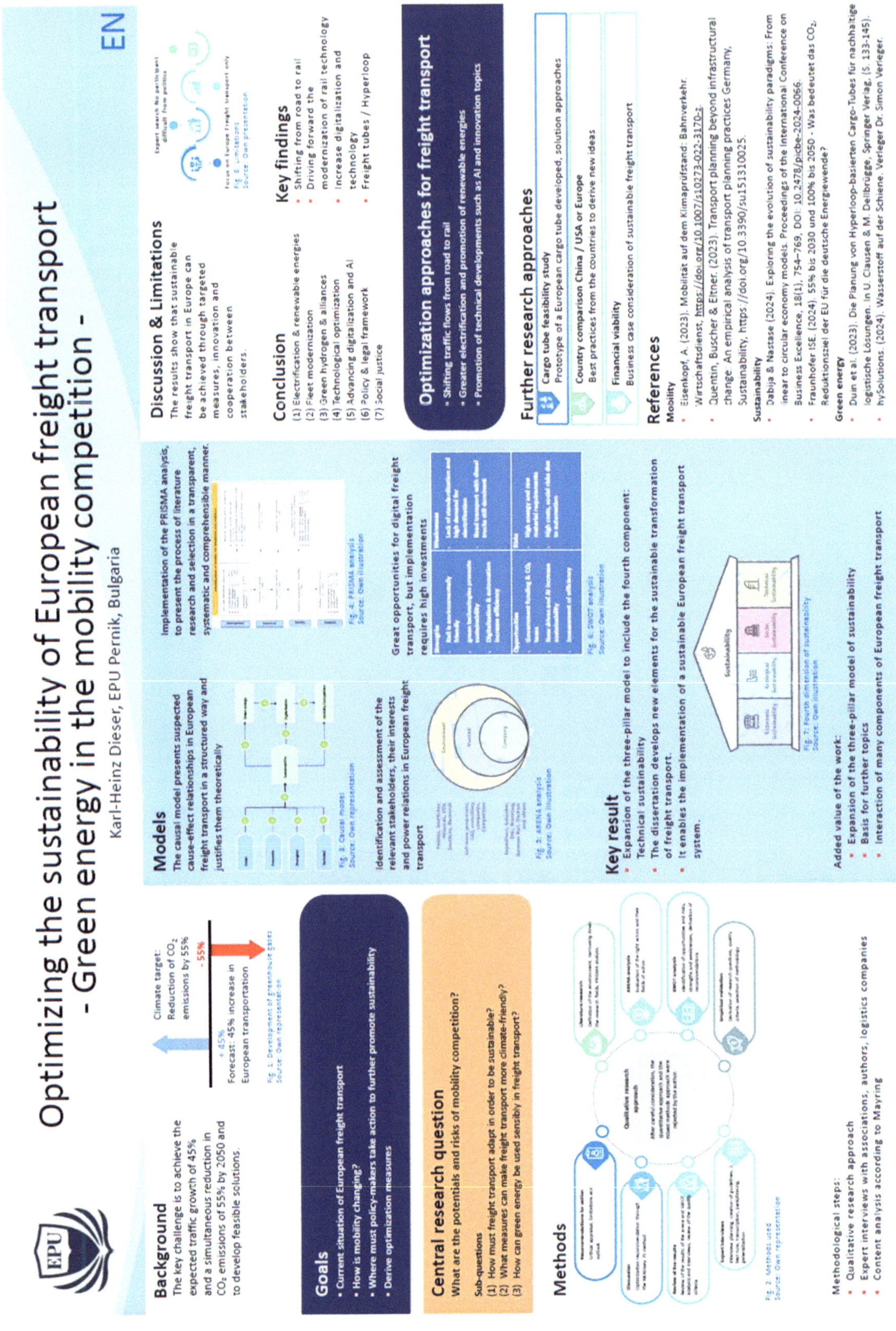

Figure 23: Research poster in English language
Source: Own representation.

Appendix 17: Research poster (German version)

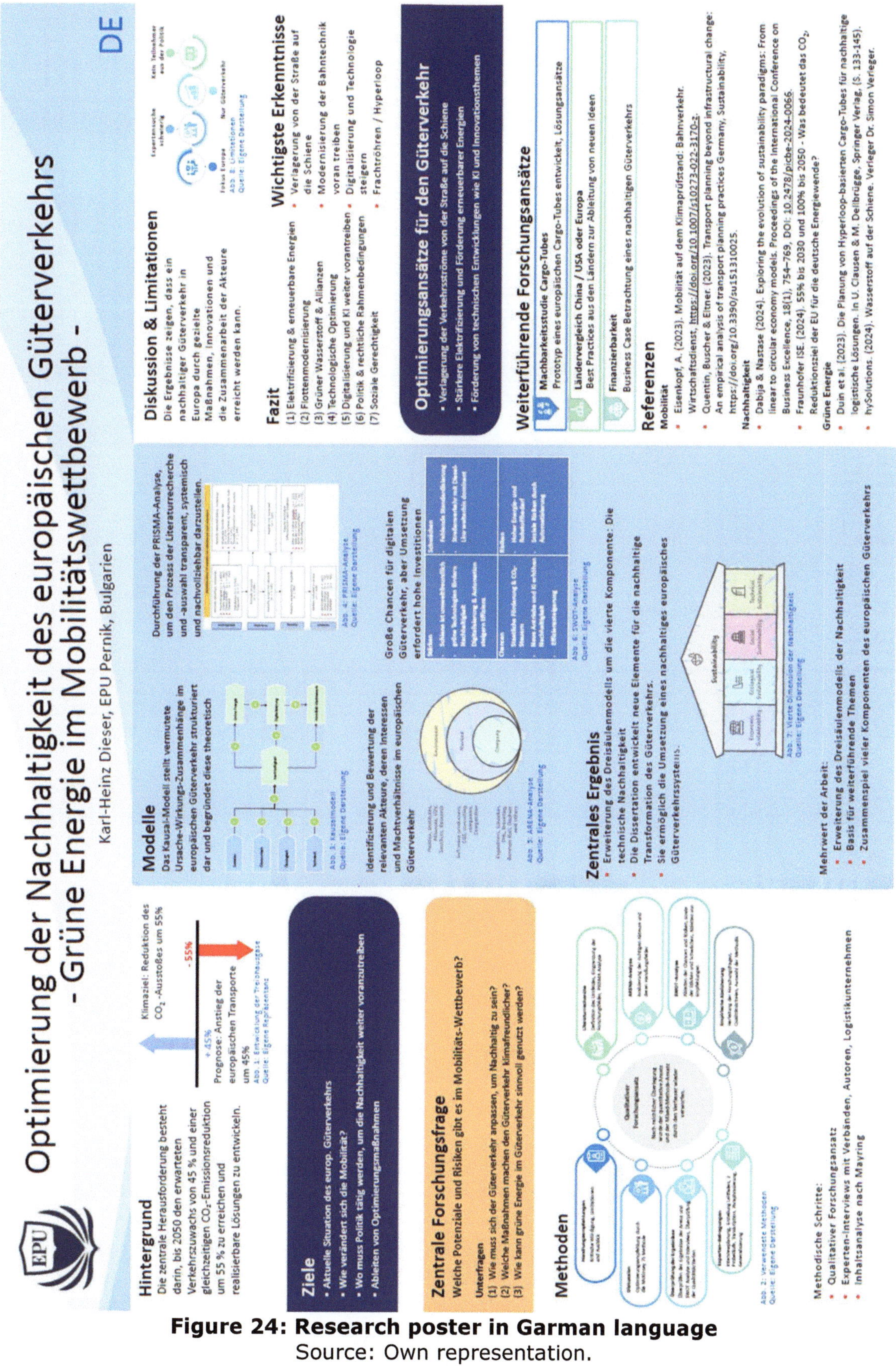

Figure 24: Research poster in Garman language
Source: Own representation.

Appendix 18: Research poster (Bulgarian version)

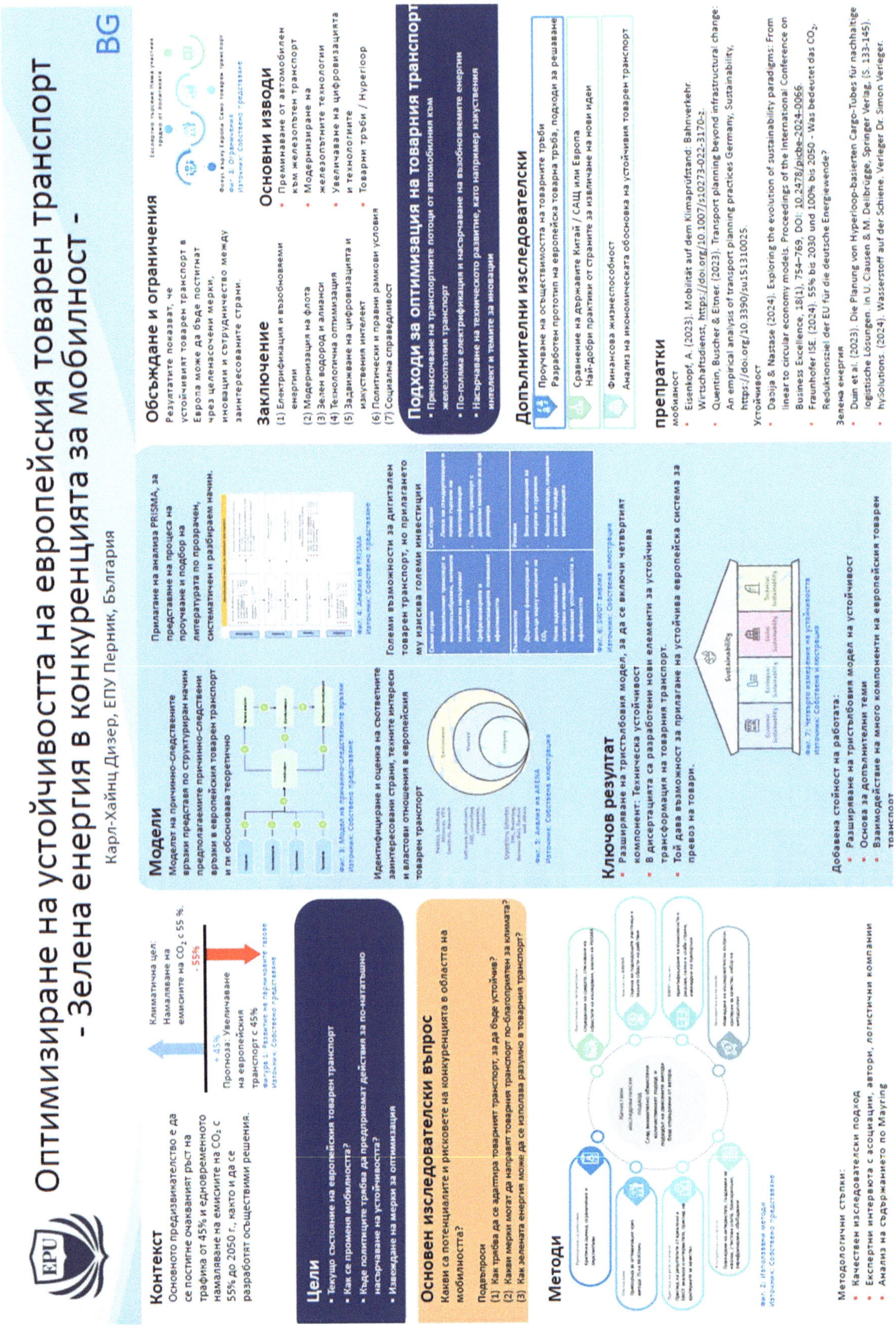

Figure 25: Research poster in Bulgarian language
Source: Own representation.

Appendix 19: Summarizing the SWOT analysis

Strengths

- Working on mobility concepts
- Cargo tubes reduce CO2 emissions and offer safe transportation
- Greening buildings reduces noise and generates oxygen
- Overhead lines bring ecological and economic benefits
- Mobility and digitalization drive European freight transport forward
- Rail is the most environmentally friendly mode of transport
- Electrification and hydrogen are gaining in importance
- Combustion engines are losing importance
- Green energy and sustainable technologies promote the future of the GF

Weaknesses

- Companies are not as green as they pretend to be
- Companies fear loss of image
- Freight transport is the third largest GHG-producing industry
- The road is still the leading mode of transport with diesel trucks
- It is still unclear which of the green technologies is the most suitable
- Rail network needs to be heavily electrified
- No standardized rail transport in Europe (track gauges, signals, etc.)
- Road transport is threatened by a shortage of drivers, high fluctuation
- Permanent traffic jams

Opportunities

- State subsidies for carbon neutral drives
- Shift from road to rail relieves pressure on infrastructure
- DAC brings digitalization to freight trains, sensor technology possible
- Hybrid drives increase sustainability
- Digitization increases efficiency
- Automation supports mobility
- E-fuels and hydrogen reduce the ecological footprint
- AI will control maintenance, loading and traffic in the future
- Progressive CO2 taxation brings economic benefits

Risks

- Electrification requires a lot of resources in the form of electrical energy
- Expansion of electrification requires large quantities of raw materials
- Mining is often harmful to the environment
- Insufficient infrastructure for alternative drive systems
- Building the infrastructure is not going fast enough
- Lack of charging structures delays
- High costs place an economic burden on companies
- New technologies and AI also entail high risks
- Autonomous systems can lead to job losses

Figure 26: Summarization of the SWOT analysis
Source: Own representation.

Appendix 20: Time schedule

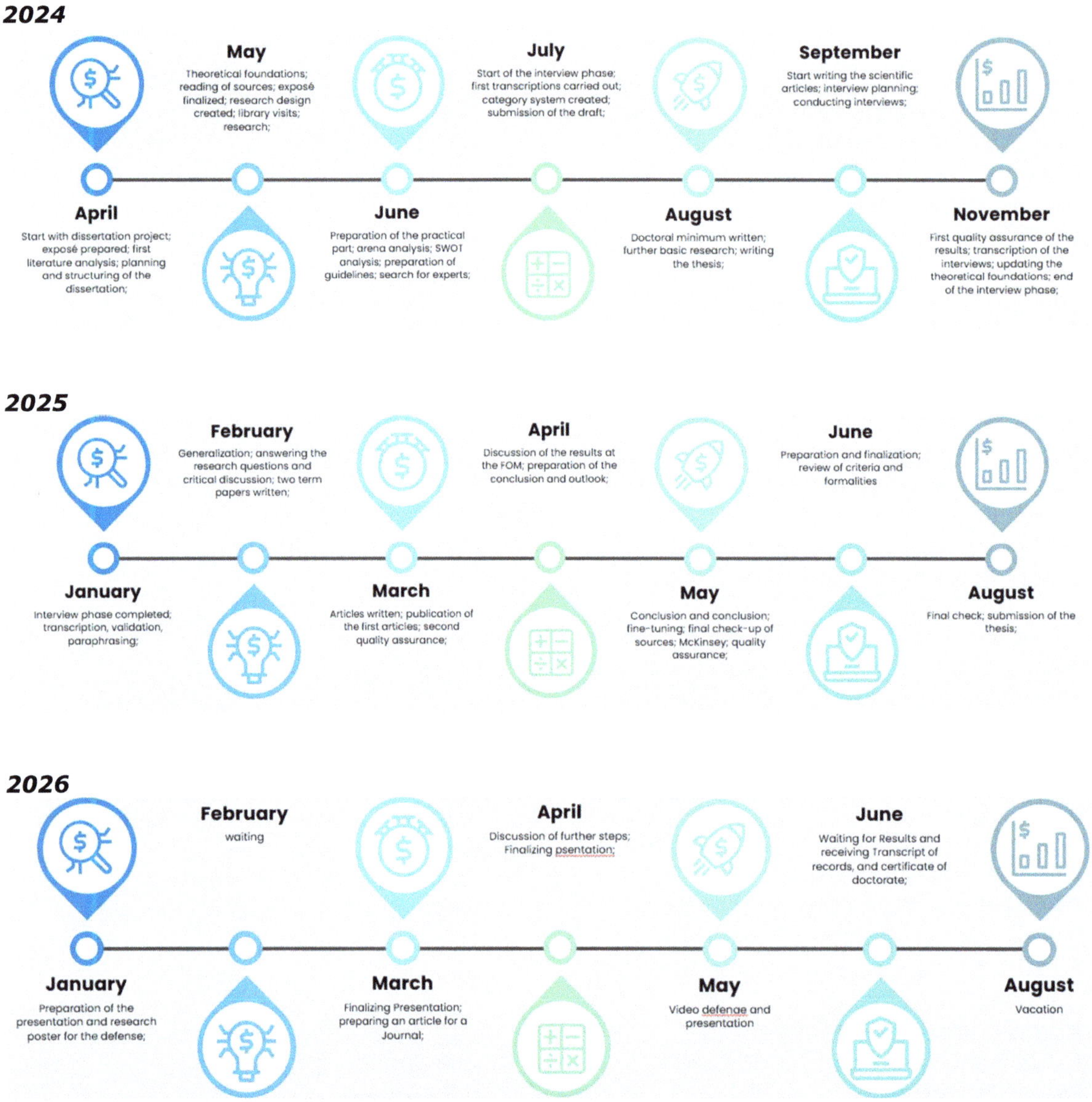

Figure 27: Time schedule
Source: Own representation.

Appendix 21: Written validation of the transcripts

Subject: Written validation of the transcripts

Dear Mr. (X),
Dear Ms. (Y),

Thank you very much for finding the time to conduct the expert interview with me a few weeks ago on the subject of "Optimizing the sustainability of European freight transport - green energy in the mobility competition".

As discussed, I am sending you the written form of the interview in the form of a transcript with the request that you proofread the transcript and, if necessary, make changes if I have not reproduced something linguistically correctly. The interview itself has been smoothed out, i.e. linguistic inconsistencies have been corrected to make it easier to read and evaluate later.

Thank you very much for your positive feedback and support! Please do not hesitate to contact me if you have any questions.

Yours sincerely,

Karl-Heinz Dieser

Appendix 22: Articles by Author

(1) Dieser, K. H. (2025). Agile project management in public administration. *European Journal of Business, Economics and Accountancy, 13 (1),* 11-22., https://www.idpublications.org/wp-content/uploads/2025/04/Full-Paper-AGILE-PROJECT-MANAGEMENT-IN-PUBLIC-ADMINISTRATION.pdf

(2) Dieser, K.-H. (2025). THE IMPORTANCE OF SMART CITIES: CHALLENGES, OPPORTUNITIES AND IMPLEMENTATION STRATEGIES [JB]. International Journal of Psychology and Strategic Communication, 02, 1–7. https://doi.org/10.61030/IJPSC.25.v02pp03

(3) Dieser, K.-H., & Frenzel, M. (2025). Effects of CO_2 certificate trading on the logistics sector in the European Economic Area: A comparative analysis of the European and German strategies. European Journal of Business, Economics and Accountancy, 13(1), 32–38. https://doi.org/10.5281/zenodo.15296928

(4) Dieser, K.-H. (2025). Transformation of the Steel Industry: Strategies for Climate-Neutral Production. International Journal of Psychology and Strategic Communication, 2(PP02), https://doi.org/10.61030/IJPSC.25.v02pp02

(5) Dieser, K. H. (2025). Applicability of the vie-model in public administration. European Journal of Psychological Research, 12 (1), 14-25, https://www.idpublications.org/wp-content/uploads/2025/05/Full-Paper-APPLICABILITY-OF-THE-VIE-MODEL-IN-PUBLIC-ADMINISTRATION.pdf

(6) Dieser, Karl-Heinz. Strategies to increase employee motivation in public administration. World Journal of Advanced Research and Reviews, 2025, 26(02), 1155-1163. Article DOI: https://doi.org/10.30574/wjarr.2025.26.2.1787, https://journalwjarr.com/sites/default/files/fulltext_pdf/WJARR-2025-1787.pdf

(7) Dieser, K. H. (2025). New public management: employee motivation in a public authority. European Journal of Research in Social Sciences, 13 (1), 24-33; https://www.idpublications.org/ejrss-vol-13-no-1-2025/

(8) Dieser Karl-Heinz. Sustainability in business: The impact of sustainable management on business success. World Journal of Advanced Research and Reviews, 2025, 26(03), 1664-1673. Article DOI: https://doi.org/10.30574/wjarr.2025.26.3.2377

(9) Dieser, K. H. (2025). Possible applications of artificial intelligence in public administration. European Journal of Engineering and Technology, 13 (1), 11-19; DOI: https://doi.org/10.5281/zenodo.16587361

(10) Dieser, K.-H. (2025). Green energy at home: Potentials and Challenges of user integration through design thinking, https://www.researchpublish.com/papers/green-energy-at-home-potentials-and-challenges-of-user-integration-through-design-thinking, https://doi.org/10.5281/zenodo.16793542

(11) Dieser, H. H. (2026). Performancesteigerung im öffentlichen Dienst. International journal of life sciences research, 14(2), 52–61. https://doi.org/10.5281/zenodo.20621830

KEYWORD INDEX

AI VIII, 13, 14, 25, 26, 30, 31, 58, 66, 81, 94, 99, 100, 106, 108, 110, 111, 112, 136, 139, 142, 162, 163, 164, 165, 184, 189, 190, 191, 202, 239
Air transport 36
causal model 60, 108
challenge XII, 29, 32, 34, 38, 40, 45, 50, 55, 63, 89, 92, 106, 107, 128, 201, 204, 219
challenges 17, 24, 27, 29, 34, 35, 41, 50, 51, 52, 53, 54, 55, 56, 80, 81, 85, 89, 90, 97, 102, 106, 109, 112, 122, 125, 200, 205, 206, 209, 212, 213, 216, 219, 232, 234
climate neutrality 14, 17, 23, 24, 39, 41, 44, 54, 205, 206, 218, 229, 235
climate neutrality targets 23
climate targets XII, 13, 14, 15, 16, 18, 19, 20, 22, 24, 40, 52, 55, 63, 64, 68, 87, 95, 98, 100, 104, 109, 196, 201, 205, 206, 208, 222, 229, 239
CO_2 emissions XII, 13, 15, 16, 19, 22, 28, 33, 34, 37, 39, 46, 47, 51, 54, 55, 63, 161, 184, 208, 215, 217, 222, 234
decarbonization 16, 22, 28, 32, 72, 73, 85, 104, 106, 112, 149, 204, 212, 215, 216, 222, 228, 232, 233, 236
digital infrastructure 14
digitalization 14, 18, 26, 29, 30, 31, 35, 40, 51, 58, 61, 74, 85, 87, 88, 90, 91, 93, 94, 99, 100, 101, 103, 104, 108, 110, 112, 143, 146, 151, 161, 164, 165, 172, 179, 188, 189, 199, 203, 211, 220, 226, 228, 231, 235
energy consumption 13, 15, 20, 30, 45, 46, 73, 88, 93, 101, 105, 131, 169, 220, 222
energy savings 17
energy sources 13, 14, 23, 33, 39, 40, 45, 60, 99, 100, 104, 105, 106, 107, 142, 153, 168, 185, 186, 192, 193, 200, 207
environmental targets 13, 30, 99, 222
EU VIII, XII, 13, 14, 15, 23, 24, 33, 37, 39, 41, 42, 43, 50, 53, 54, 56, 110, 114, 116, 117, 119, 120, 121, 122, 124, 130, 131, 132, 133, 134, 135, 142, 193, 199, 202, 203, 205, 206, 207, 208, 209, 211, 216, 220, 221, 223, 224, 228, 229, 230, 236, 237
European Green Deal 17, 22, 23, 24, 39, 41, 54, 56, 58, 196, 205, 207, 221, 222, 229, 236
European railroads 17
European targets 14, 15
expert interviews 18, 20, 21, 58, 59, 60, 62, 66, 69, 70, 77, 78, 79, 82, 83, 84, 87, 88, 96, 98, 107, 108, 109, 138, 141
freight transport I, III, XII, 13, 14, 15, 16, 17, 18, 19, 20, 21, 22, 29, 31, 33, 34, 35, 36, 37, 40, 41, 43, 44, 46, 47, 48, 49, 50, 51, 52, 53, 54, 55, 57, 58, 63, 64, 65, 66, 67, 68, 69, 70, 73, 74, 75, 77, 79, 80, 81, 82, 84, 85, 87, 88, 89, 90, 91, 92, 93, 94, 95, 96, 97, 98, 99, 100, 101, 102, 103, 104, 105, 106, 107, 108, 109, 111, 112, 113, 115, 121, 122, 126, 127, 131, 138, 139, 148, 152, 158, 160, 168, 173, 175, 179, 183, 185, 200, 201, 202, 204, 206, 207, 208, 211, 213, 214, 215, 217, 219, 220, 222, 223, 224, 226, 228, 229, 230, 232, 234, 235, 237, 239, 245
Freight transport XII, 13, 15, 18, 33, 35, 37, 40, 41, 48, 49, 51, 54, 85, 144, 146, 159, 173, 178, 188, 201, 207, 210, 213, 220, 222, 223, 227, 229, 232, 236, 237
fuel cells are 39, 89
globalization 29
green energy 13, 14, 17, 19, 20, 22, 37, 55, 58, 61, 62, 64, 65, 66, 69, 70, 75, 82, 94, 95, 106, 139, 166, 191, 193, 206, 216, 229, 234, 239, 245
Green energy 17, 19, 22, 23, 39, 40, 60, 64, 143, 205, 222, 226, 237
Green Energy I, III, 22, 62, 74, 75, 87, 118, 128, 133, 138, 139, 142, 143, 166, 196, 206, 222, 228
greenhouse gas emissions 13, 23, 24, 33, 39, 40, 41, 47, 85, 95, 167, 202, 222
greenhouse gases 13, 22, 23, 33, 34, 95, 169
greenwashing 17, 85, 91, 103, 132, 147, 157, 179, 180, 219
hydro power 17
hydrogen XII, 14, 18, 34, 38, 39, 51, 54, 55, 56, 89, 90, 92, 99, 100, 102, 103, 104, 105, 107, 109, 110, 112, 124, 127, 129, 131, 143, 145, 146, 149, 151, 163, 170, 173, 177, 178, 185, 193, 203, 204, 208, 212, 215, 217, 219, 220, 223, 228, 230, 233, 234, 236, 239
Inland navigation 36
inland waterways 18, 35, 48, 49, 223
Innovative 14, 39, 90, 126, 214
kilowatt 45
logistics 13, 18, 28, 31, 33, 37, 38, 39, 40, 45, 47, 48, 52, 55, 58, 63, 67, 68, 79, 81, 85, 92, 93, 97, 100, 101, 103, 104, 110, 112, 122, 125, 127, 129, 131, 142, 170, 182, 184, 196, 199, 200, 201, 202, 208, 210, 211, 212, 215, 216, 217, 218, 219, 220, 224, 226, 227, 230, 231, 233, 234, 235, 236, 237
logistics concepts 13, 224
low-emission drive systems 40
management systems 17, 34, 223
Maritime shipping 36, 43
megajoules 45
mobility I, III, 13, 14, 15, 16, 17, 18, 19, 20, 22, 28, 31, 32, 33, 34, 38, 39, 40,

44, 51, 53, 57, 58, 60, 61, 62, 63, 66, 74, 81, 82, 85, 98, 101, 102, 104, 107, 109, 110, 114, 117, 122, 125, 132, 138, 142, 152, 172, 175, 196, 199, 201, 202, 203, 208, 209, 210, 211, 212, 213, 214, 216, 218, 222, 224, 226, 227, 230, 231, 232, 233, 236, 237, 239, 245
Mobility 31, 32, 33, 34, 39, 40, 43, 51, 62, 75, 85, 132, 142, 172, 175, 182, 187, 196, 203, 204, 208, 214, 216, 218, 219, 220, 223, 226, 228, 230, 232, 234, 235, 237, 239
Paris Climate Agreement 17, 22, 39, 222, 236
pillar of sustainability 26
Pipework 36
PRISMA analysis 58
Rail transport 36, 48, 223, 237
reduce emissions XII, 17, 28, 33, 54, 56, 87, 99, 105, 235
regional fluctuations 17
renewable XII, 13, 14, 15, 16, 17, 22, 23, 26, 27, 33, 38, 39, 40, 47, 50, 51, 55, 60, 62, 64, 90, 95, 99, 100, 103, 104, 110, 112, 116, 128, 143, 149, 175, 185, 186, 200, 202, 205, 207, 213, 215, 216, 218, 221, 222, 233, 236
Renewable energies 46, 147, 175, 178, 196, 218, 235
renewable energy XII, 13, 14, 15, 16, 17, 22, 23, 33, 39, 40, 47, 50, 55, 60, 62, 64, 90, 99, 100, 103, 104, 128, 175, 185, 186, 200, 205, 207, 215, 221, 222, 233, 236
Road 15, 33, 36, 44, 46, 49, 50, 98, 132, 148, 175, 215, 222
Strategy 98, 200, 239
Structure 20, 98
sustainability I, III, XII, 13, 15, 16, 17, 18, 19, 20, 22, 24, 25, 26, 27, 28, 29, 30, 31, 40, 47, 48, 51, 53, 55, 56, 58, 60, 61, 63, 64, 65, 66, 67, 69, 70, 72, 73, 74, 75, 76, 81, 85, 87, 88, 89, 90, 91, 92, 93, 94, 95, 96, 98, 99, 100, 102, 103, 104, 105, 106, 107, 108, 109, 110, 111, 115, 116, 118, 122, 126, 128, 130, 138, 139, 142, 147, 148, 149, 152, 153, 157, 161, 167, 169, 172, 173, 175, 178, 179, 183, 184, 185, 190, 192, 196, 199, 201, 202, 204, 205, 209, 211, 213, 214, 216, 218, 220, 222, 227, 230, 232, 233, 235, 236, 239, 245
Sustainability VIII, 14, 25, 26, 27, 33, 37, 40, 53, 60, 62, 73, 74, 107, 114, 115, 122, 126, 128, 129, 142, 143, 145, 146, 147, 173, 176, 179, 185, 192, 200, 201, 203, 209, 210, 211, 212, 214, 215, 216, 224, 227, 228, 231, 233, 237, 239
Sustainability strategies 33
Sustainable energy 18
SWOT analysis 18, 20, 21, 58, 59, 62, 78, 80, 84, 86, 98, 100, 109, 140, 240, 241, 242, 243, 244
SWOT matrix 59
Systems 98, 115, 116, 122, 124, 130, 202, 210
transport infrastructure 34, 35, 37, 92, 93, 96, 100, 105, 106
transportation system 34, 44, 47, 50, 202, 227
transportation systems 15, 24, 31, 40, 53, 85
work cultures 30